Grifting The Amazon

Angie Waller

Introduction

In 2003, I published *Data Mining the Amazon,* my first book. Its contents were inspired by the creepy feeling I got when Amazon's recommendation algorithm suggested I might like to listen to the rock band Stone Temple Pilots since I had purchased a book about web programming. *Data Mining the Amazon* was designed to resemble an annual report, but instead of financial figures, I reverse engineered the company's recommendation system for books and CDs. It was a paperback book printed in the Midwest at an offset press, distributed as an art project, and exhibited in galleries and museums. I also sent it to people who purchased it from my website for $15.

I found publishing a book with an offset press to be financially unsustainable. With *Data Mining the Amazon*, I spent over a thousand of my student loan dollars to print 500 books. To distribute copies, I found myself standing in long lines at the post office in order to fill out customs forms for international orders. The large inventory of books packed in heavy boxes was also a burden as I moved from studio to studio.

The print-on-demand industry eased these financial and logistical problems. I was able to print one book for less than $10 instead of shelling out thousands of dollars at a time. In addition, self-publishing companies like CreateSpace (now owned by Amazon) made it possible to print and distribute books through Amazon.com. I no longer had to worry about storage space and spending my lunch breaks in long post office lines.

This book documents my experience of engaging with Amazon.com to publish and distribute my work. While most of the action in this recounting involves watching online videos, uploading files, and waiting for UPS, a colorful cast of real people are behind the screen. Along the way I encountered a Danish male fashion model, a prolific ghostwriter from India, and a "tight" community of Australian capybara pet owners. I even witnessed my own books in potentially criminal hands.

The title of this book intentionally mirrors the title of my first book; however, I am no longer just an observer of the platform. *Grifting the Amazon* explores the metalanguage around producing books in the tangled world of e-commerce on Amazon.com.

The Medium is the Message

Books are products

One day, when I was feeling particularly unemployed and restless, I found myself in my kitchen listening to the soothing English accent of a Danish male fashion model whom I will refer to as NK. This particular morning, I had decided to switch my focus from job hunting to learning how to publish on Amazon Kindle. That desire led me to NK. He wasn't sitting in my kitchen, but from my computer his disembodied voice narrated a series of screen capture video tutorials for publishing Kindle eBooks. Admittedly, having a Skillshare.com course taught by a male model—even one that never showed his face—seemed a lot cooler than the other programming and design classes offered on the website.

The class was ostensibly about designing and formatting text for Kindle. However, the class barely covered technical specifications or related software. Instead, NK demonstrated how to get books on Amazon.com without writing a word or designing a single page.

> “Amazon has over 600 million credit cards on file. That’s a lot of credit cards. So the buying audience is so abundant it’s not even funny.”
>
> –NK, full time model and Kindle publisher

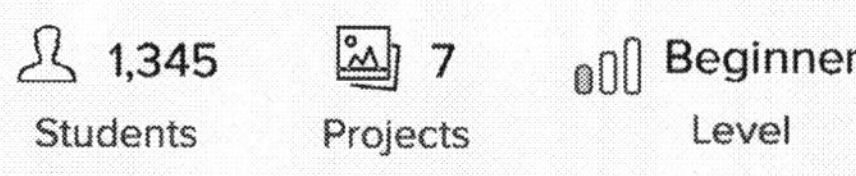

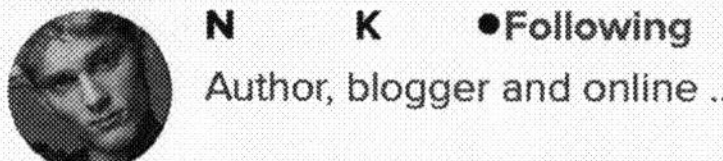

About Me

I'm a full time model who decided I wanted to establish an additional source of income. After playing around with different ways of building an online source of passive cash flow, and giving up on every single one of them, I finally discovered the Amazon Kindle platform. I found out how relatively simple it was to make money providing quality content, while still keeping a day job and a social life.

One thing I want to make clear is this requires work, not as much as other methods, but it is definitely not a get rich quick scheme. With spending a few hours each week I established enough passive income to cover my monthly groceries expenses with six short eBooks on the Kindle store.

I'm no Kindle publishing wizard making six figures on this, in fact just months ago I knew nothing of this, but I now have a small steady source of income that I can scale into something bigger down the line - which I will. I have now grown my passive income to +$1500 a month.

I'm a blonde male model from Denmark - if I can do this, you certainly can too.

↑ Introduction to NK’s Skillshare.com class.

Author Anonymity

Lifestyle brand as author

At the time I watched the tutorial in 2015, self-published paperbacks and Kindle eBooks were being published on Amazon.com through a website called CreateSpace, now owned by Amazon and called Kindle Direct Publishing (kdp.amazon.com). Making a book on CreateSpace simply entailed uploading PDFs of your book's contents and cover.

CreateSpace had few constraints on how a book was titled or authored. There was no verification that the name of the author was connected to a real identity—any name or fictional brand could be listed as the book's author. For instance, NK calls his brand "Sound and Simple Lifestyle."

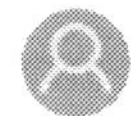

Sarah

I read the entire book in 8 minutes and learned ...

November 24, 2017

Format: Kindle Edition | **Verified Purchase**

I read the entire book in 8 minutes and learned absolutely nothing. The grammatical band syntax errors were, however, amusing.
No wonder there is no author named, it is an embarrassment

Helpful | Comment | Report abuse

↑ Review for a "Sound and Simple Lifestyle" book commenting on the author's anonymity.

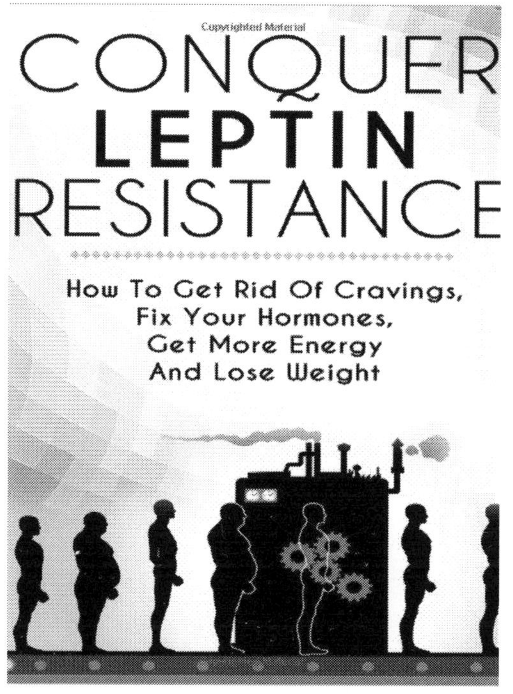

+ Follow

Follow to get new release updates and improved recommendations

About Sound And Simple Lifestyle

Sound & Simple Lifestyle offers you ebooks on making the most out of the hand you've been dealt. As the name implies we provide sound and simple steps to leading a healthier and more productive life, whether that being through cooking, exercise etc. We hope you enjoy our material and that they will help you be the best you.

↑ Example of a book published on Amazon.com under NK's brand "Sound and Simple Lifestyle."

↑ "About Sound and Simple Lifestyle" from Amazon.com.

Product details

Paperback: 56 pages
Publisher: CreateSpace Independent Publishing Platform (October 21, 2013)
Language: English
ISBN-10: 1495341968
ISBN-13: 978-1495341960
Product Dimensions: 8.5 x 0.2 x 11 inches
Shipping Weight: 7 ounces (View shipping rates and policies)
Average Customer Review: ★★★☆☆ 13 customer reviews
Amazon Best Sellers Rank: #4,040,959 in Books (See Top 100 in Books)
#378095 in Health, Fitness & Dieting (Books)

↑ Details for a "Sound and Simple Lifestyle" book published on CreateSpace.com. CreateSpace Independent Publishing Platform is listed as the publisher.

Titles Come Before Content

The more keywords the better

To ensure that a book topic would be niche enough for success, NK demonstrated how to fine-tune book titles using Google's AdWords tool. With AdWords (now known as Google Ads), you can determine how many people are searching for a subject online and the keywords they are using to access the information. For instance, instead of just using the word "pickling" to title a book about pickling vegetables, AdWords might also recommend a related term like "fermenting." If there are fewer books with the words "fermenting" *and* "pickling" in the title compared to books with just "pickling," you could rank higher on Amazon.com and potentially gain more sales. These competitive keywords have inspired the contents of NK's books.

"An e-book on Amazon can sell for at least just as much as an iPhone app, but it can be produced at a fraction of the price. This is why Kindle eBook publishing is the simplest and the most replicable online income opportunity out there."

–NK, full time model and Kindle publisher

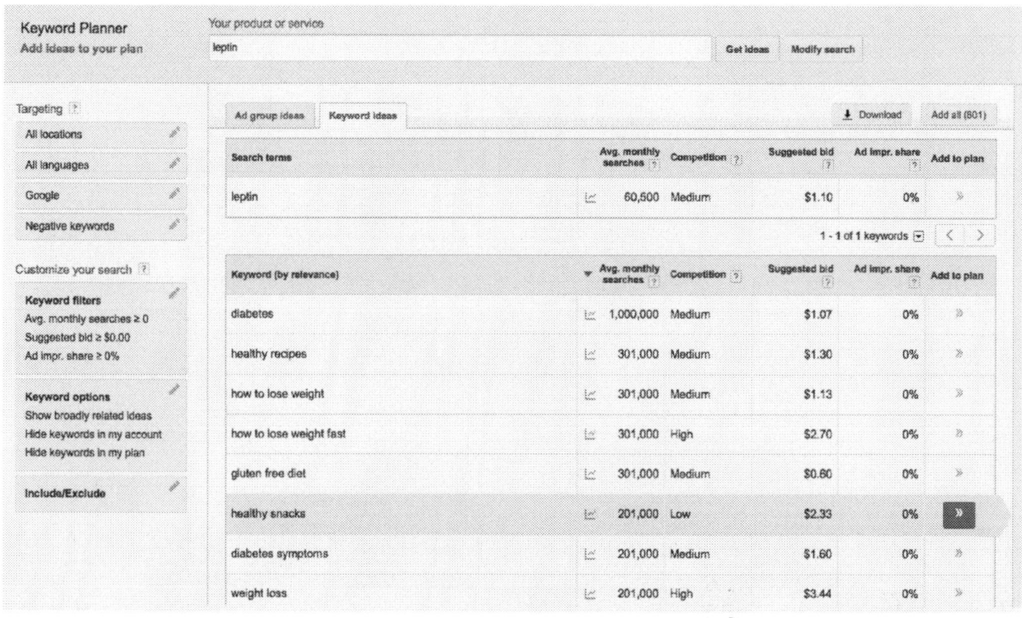

Search terms	Avg. monthly searches	Competition	Suggested bid	Ad impr. share	Add to plan
leptin	60,500	Medium	$1.10	0%	»

Keyword (by relevance)	Avg. monthly searches	Competition	Suggested bid	Ad impr. share	Add to plan
diabetes	1,000,000	Medium	$1.07	0%	»
healthy recipes	301,000	Medium	$1.30	0%	»
how to lose weight	301,000	Medium	$1.13	0%	»
how to lose weight fast	301,000	High	$2.70	0%	»
gluten free diet	301,000	Medium	$0.60	0%	»
healthy snacks	201,000	Low	$2.33	0%	»
diabetes symptoms	201,000	Medium	$1.60	0%	»
weight loss	201,000	High	$3.44	0%	»

↑ Still from NK's Skillshare.com tutorial "How to Title Your Book." The table is from Google's AdWords tool.

↓ Publications from "Sound and Simple Lifestyle" that use keywords from NK's Google AdWords research.

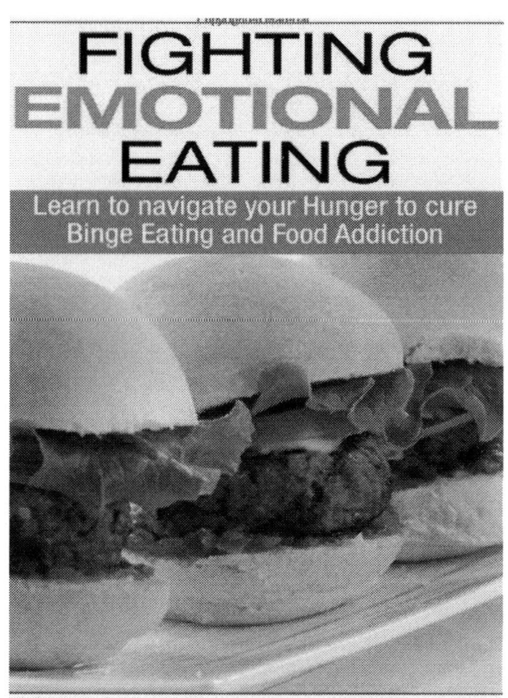

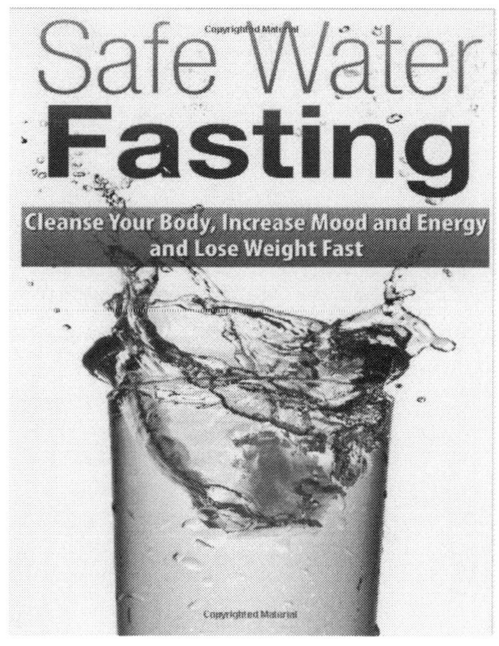

Writing a Book

Getting someone else to write a book for you

The most time-intensive part of NK's process must have been selecting his title. The rest of the production, including cover design and content writing, was outsourced overseas. Being an international male model like NK is not a requirement for meeting writers and designers in faraway places. Websites like Upwork.com (formerly Elance.com) and Fiverr.com are venues for posting short job descriptions and accepting bids from writers and designers all around the world.

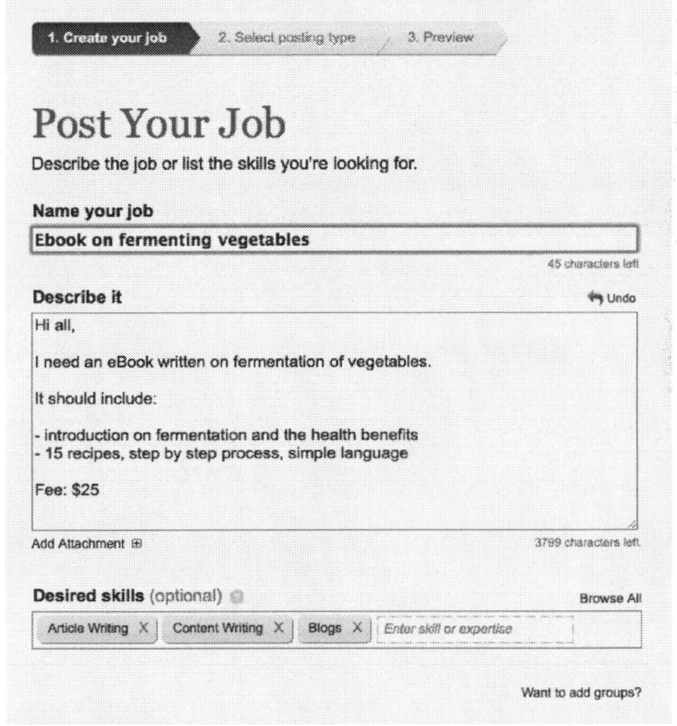

↑ Screen capture from NK's tutorial on hiring a ghostwriter on Elance.com.

> "...Now I am outranking these books and I'm just a dude who hired a medical student in Pakistan to write my book. So you see it is more about how you package the book rather than the content itself."
>
> –NK, full time model and Kindle publisher

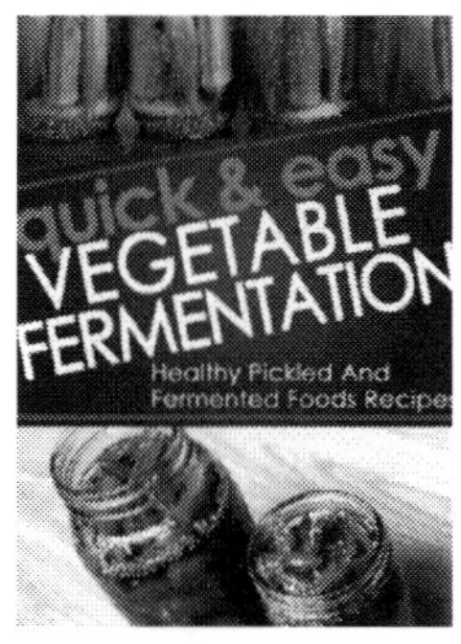

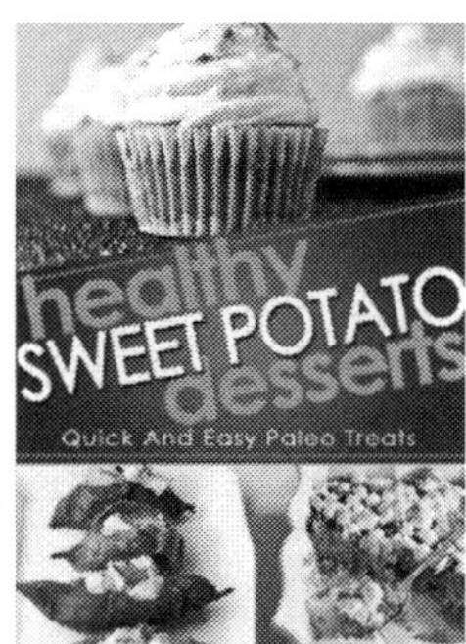

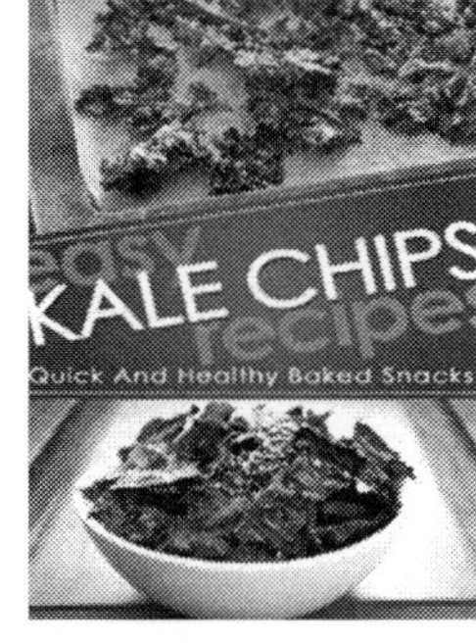

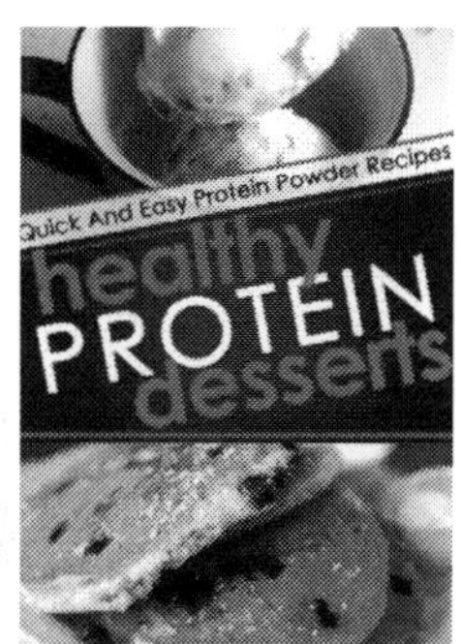

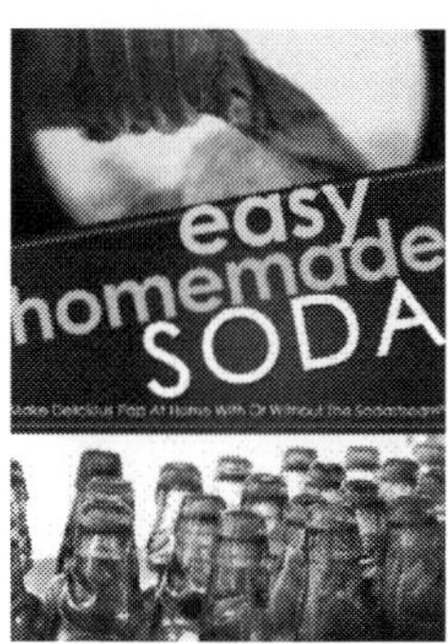

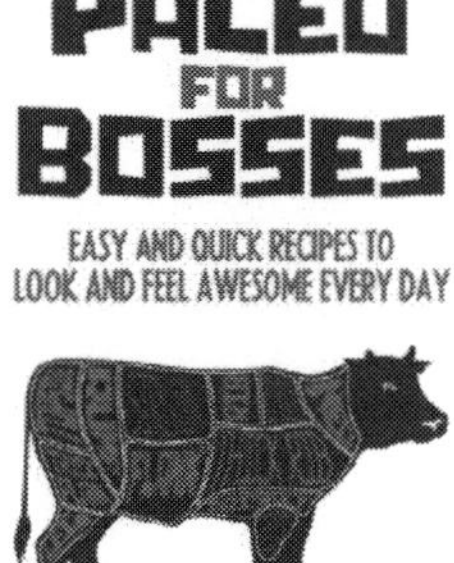

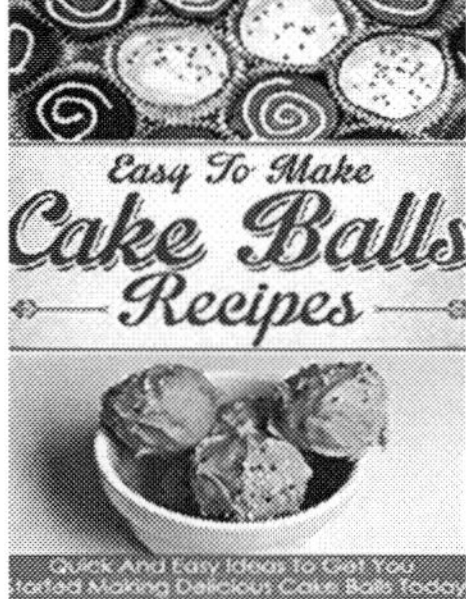

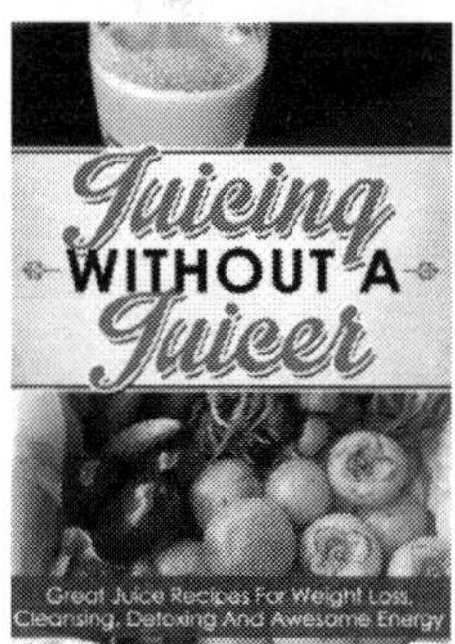

↑ A selection of titles by "Sound and Simple Lifestyle." Each cover was designed by a freelancer on Fiverr.com, another website for finding freelance talent for small gigs.

Ghostwriter Job Posts

Tight timeline and minimal guidance

Upwork.com is filled with listings that reveal a variety of approaches to asking internet strangers to churn out books. Some job posters want ongoing relationships with the same writer, while others have vague themes in mind, word count requirements, and a tight turnaround time. Some only provide word count and want ghostwriters to propose their own book topics.

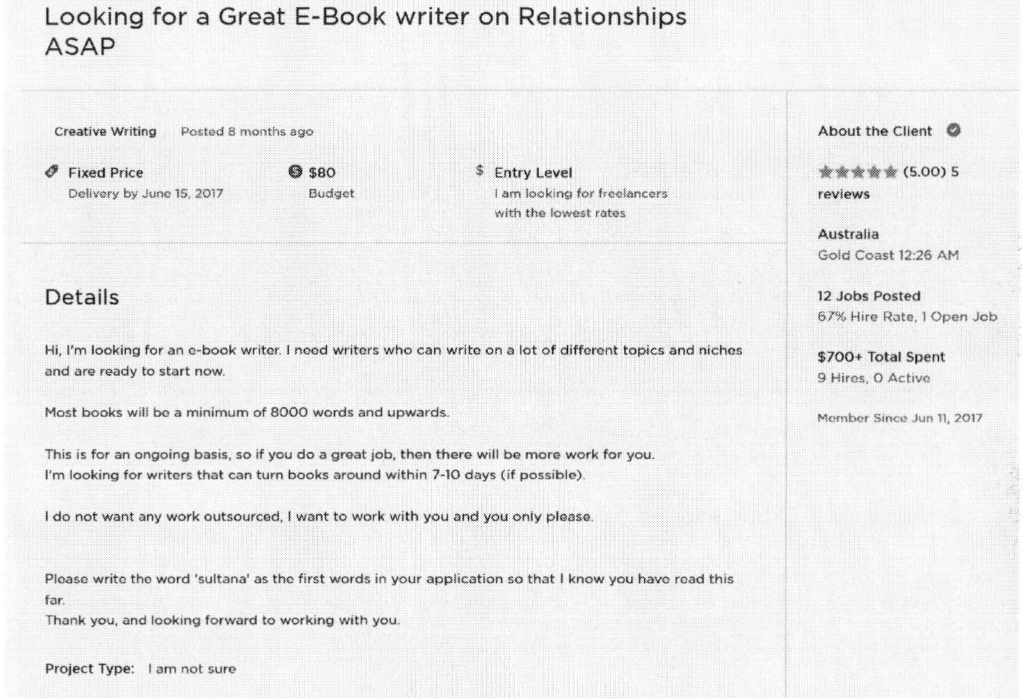

Looking for a Great E-Book writer on Relationships ASAP

Creative Writing Posted 8 months ago

Fixed Price
Delivery by June 15, 2017

$80
Budget

Entry Level
I am looking for freelancers with the lowest rates

Details

Hi, I'm looking for an e-book writer. I need writers who can write on a lot of different topics and niches and are ready to start now.

Most books will be a minimum of 8000 words and upwards.

This is for an ongoing basis, so if you do a great job, then there will be more work for you.
I'm looking for writers that can turn books around within 7-10 days (if possible).

I do not want any work outsourced, I want to work with you and you only please.

Please write the word 'sultana' as the first words in your application so that I know you have read this far.
Thank you, and looking forward to working with you.

Project Type: I am not sure

About the Client

(5.00) 5 reviews

Australia
Gold Coast 12:26 AM

12 Jobs Posted
67% Hire Rate, 1 Open Job

$700+ Total Spent
9 Hires, 0 Active

Member Since Jun 11, 2017

↑ Screen capture from Upwork.com documenting a client looking to establish an ongoing relationship with a ghostwriter who will compose 8,000-word books on a weekly basis for $80, or one cent per word.

↓ Screen capture from Upwork.com showing a request for 5500-word books on a topic and title proposed by the ghostwriter.

Details

Hi

I'm looking for a writer who has experience in writing eBooks/Kindle Books.

I don't have a title for the book because I want you to propose what you are able to write.

The book must consist of 5,500 words and must pass copyscape.

Interested applicants please send me the following and I will get back to you if I am keen to work with you.

1. What is your proposed book topic/title?
2. What makes you qualified to write this topic?
3. Have you written on a similar topic before? if yes, please send me some references. Thank you.

Interested applicants please apply.

P.S. Only applicants who answered the above three questions will be considered. Thank you.

Project type: One-Time Project

★☆☆☆☆ Be careful with this client: they tell you you've stolen information when you've clearly cited it, provide short deadlines then refuse to communicate back for days at a time, delete milestones and then refuse to acknowledge/add them back so that you can progress working on the contract, are rude (very much so sugar and ice personality-- when you have any questions, concerns, or need clarifications, immediately become hostile), and refuse to provide any real type of framework/direction. less

Oct 2017 - Nov 2017
Fixed Price $5.00

To Freelancer: K No feedback given

I need a ghostwriter for 3500 word e-book, buddhism topic

No feedback given

Aug 2017 - Sep 2017
Fixed Price $11.00

To Freelancer: M ★★☆☆☆

↑ Screen capture from Upwork.com showing review of a client by a ghostwriter who had a negative experience.

Hiring a Ghostwriter

A new type of automatic writing

I completed all of NK's tutorial videos in a day, and perhaps this compressed timeline amplified their brainwashing effect. That evening, I hired three ghostwriters to see if the transaction was as easy as NK demonstrated. I initiated the process as if it were a form of automatic writing. I provided an outline to the ghostwriters, and the document I received back became the book. I didn't ask the ghostwriter for edits or check the finished work for plagiarism.

My books were inspired by search queries collected through a marketing tool called Keyword Discovery (keyworddiscovery.com). This tool provides the option to analyze Google searches that are in the form of full sentences. My titles based on these long phrases were niche, but not necessarily "competitive" since they were searches collected from only a few individuals.

→ Samples of front and back covers for books I published using the techniques from NK's Skillshare.com class.

How to Find a Friend, Be a Good Friend, and Delete a Friend

unknown
UNKNOWNS

↑ Front cover, *How to Find a Friend, Be a Good Friend, and Delete a Friend*, 2015.

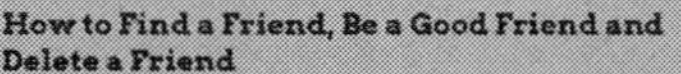
How to Find a Friend, Be a Good Friend and Delete a Friend

This volume in the Unknown Unknowns self-help series was written by a ghostwriter based in Bhubaneswar, India. S/he has written about numerous topics including: self development, relationships, travel, personal finance, business, wellness, holistic living, beauty, fashion, health, aviation and home improvement.

For more books in the series, visit unknownunknowns.org.

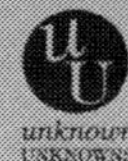

↑ Back cover, *How to Find a Friend, Be a Good Friend, and Delete a Friend*, 2015.

How to Tell if Someone is Listening to You in Your House

unknown
UNKNOWNS

↑ Front cover, *How to Tell if Someone is Listening to You in Your House*, 2015.

How to Tell if Someone is Listening to You in Your House

This book was written by a ghostwriter based in Mumbai, India. S/he has written memoirs, wine tasting guides and romance novels.

For more books in the series, visit unknownunknowns.org.

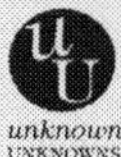

↑ Back cover, *How to Tell if Someone is Listening to You in Your House*, 2015.

Ghostwriting Life

Writing 400 books in four years

A ghostwriter I worked with, whom I will call MS, started my book on a Tuesday and by Friday it was complete (despite my flexible deadline). The book was filled with advice about friendship without any citations. Facts were editorialized with bittersweet excerpts like, *"Your parents may disinherit you tomorrow... but TRUE friends will always be there for you..."*

It seemed appropriate that the title I chose was originally a search query and the contents of the book were an eclectic collection of random blog posts, Wikipedia entries, and personal flourishes. I referred to my series of ghostwritten books as "artisanal search results."

Over email, I interviewed MS about her process (and paid her for her time). I learned that she wrote more than 400 books in four years, ranging from 5,000 to 15,000 words. She specializes in the "pet manual" genre, and these books are even longer, averaging about 35,000 words. Based in Bhubaneswar, India, MS works from home. Between doing chores and caring for her son, she writes e-books and manages clients on the internet for about ten hours a day. MS preferred ghostwriting to her previous job as a corporate trainer, which required her to travel and be away from her family extensively.

With sentences like *"the objective of this chapter is to lead you through various ways in which you can make friends,"* MS's corporate communications background showed through in her commissioned work.

Work history and feedback

Newest first

Pac Fr book Nov 2018 - Aug 2019 No feedback given	**$330.00** Fixed-price
Professional diets cookbook writers are wanted!!! HIGH BUDGET!! Jul 2018 - Jul 2019 No feedback given	**$75.00** Fixed-price
Ebook for Meditation for beginners/destress/working class Jul 2018 - Nov 2018 No feedback given	**$70.00** Fixed-price
Hissing C ★★★★★ **5.00** Oct 2018 - Nov 2018 *Great freelancer to work with.*	**$330.00** Fixed-price
A S T pets ★★★★★ **5.00** Aug 2018 - Sep 2018 *Great freelancer to work with.*	**$330.00** Fixed-price

↑ Reviews for MS's work from clients on Upwork.com.

"Usually my books do not require edits since I take great care to proofread them in detail .... However, I am open to any number of edits if the client so desires. I firmly believe that the client must get what he/she desires since I am being paid to do the job."

–MS, Ghostwriter based in Bhubaneswar, India

Chapter 1

What is Friendship?

It would be a futile attempt to define friendship at one go. It's just like any other larger than life concepts (such as romance and trust) that you cannot put into words. However, for the purpose of understanding, friendship is a bond that exists between two people that binds them together. The definition is obviously lacking in various aspects but you cannot expect to learn about friendship from a book.

Friendship has essentially two parties, each being called a friend. The relationship of friendship comes without obligations and duties and often relies on impulse and honesty. It also requires people to have each other's backs in times of need.

Friends are said to be the best gift one may find in the world. At the risk of sounding overtly cheesy, it is said that friends are there for life. Your parents may disinherit

7

↑ Excerpt from book, *How to Find a Friend, How to Be a Good Friend, and Delete a Friend*, 2015 (written by MS).

Chapter 4

Role of the Internet in Forming Friendships

Today the world has taken giant leaps into the realms of technology, which places at man's disposal an array of luxuries. Online shopping has slowly replaced a walk down to the supermarket during weekends. Internet is a giant phenomenon for humankind; so much so that from a mere pin to a humongous piano comes in a range of colors and shapes, all at a mere click of your mouse.

This comes as a blessing when it comes to finding a companion. It does not necessarily begin from the intention of hunting company though. For the purposes of socialization you can open a profile at any of the famed social networking sites.

Some of the common ones are Facebook, twitter, tumblr, LinkedIn, Pinterest, Google Plus, Instagram,

19

↑ Excerpt from book, *How to Find a Friend, How to Be a Good Friend, and Delete a Friend*, 2015 (written by MS).

Shady Pet Guru

A pet manual for every species

Intrigued by the large bounty of pet manuals she produced, I followed a link to a work sample MS provided. One book selling on Amazon.com is about taking care of capybaras and a persona named "Lolly Brown" is listed as the author. The capybara book is only a small part of the large oeuvre on exotic pets credited to Brown (see pgs. 22-23).

Brown's author biography on Amazon.com tells of owning a 120-gallon aquarium when she was a kid, which led to her "go-big-or-go-home mentality." However, an actual capybara owner who read Brown's book, *Capybara: A Complete Owner's Guide*, was unimpressed with the shallow level of Google research that went into the book's contents.

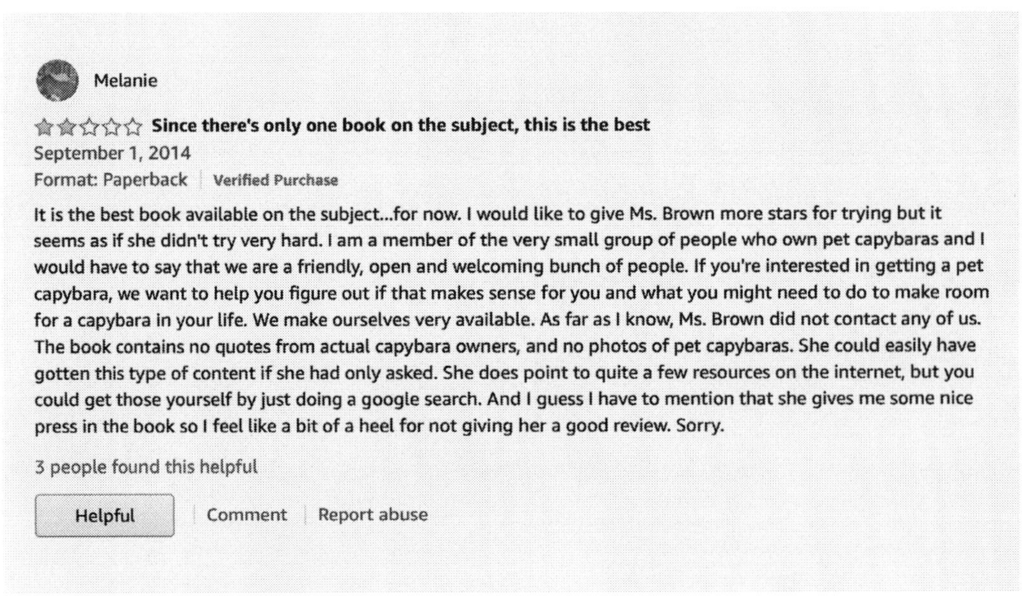

Melanie

Since there's only one book on the subject, this is the best

September 1, 2014

Format: Paperback | Verified Purchase

It is the best book available on the subject...for now. I would like to give Ms. Brown more stars for trying but it seems as if she didn't try very hard. I am a member of the very small group of people who own pet capybaras and I would have to say that we are a friendly, open and welcoming bunch of people. If you're interested in getting a pet capybara, we want to help you figure out if that makes sense for you and what you might need to do to make room for a capybara in your life. We make ourselves very available. As far as I know, Ms. Brown did not contact any of us. The book contains no quotes from actual capybara owners, and no photos of pet capybaras. She could easily have gotten this type of content if she had only asked. She does point to quite a few resources on the internet, but you could get those yourself by just doing a google search. And I guess I have to mention that she gives me some nice press in the book so I feel like a bit of a heel for not giving her a good review. Sorry.

3 people found this helpful

Helpful | Comment | Report abuse

↑ Review for Lolly Brown's book on capybara pets questioning Brown's expertise.

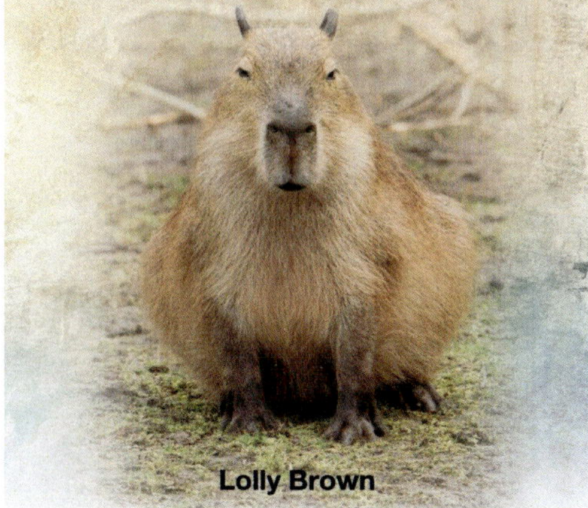

Capybara. Facts & Information: Habitat, Diet, Health, Breeding, Care, and Much More All Covered. A Complete Owner's Guide, an 86-page book published by Lolly Brown in March 2014.

Amazon.com biography of Lolly Brown, pseudonym for an author who hires ghost-writers on Upwork.com.

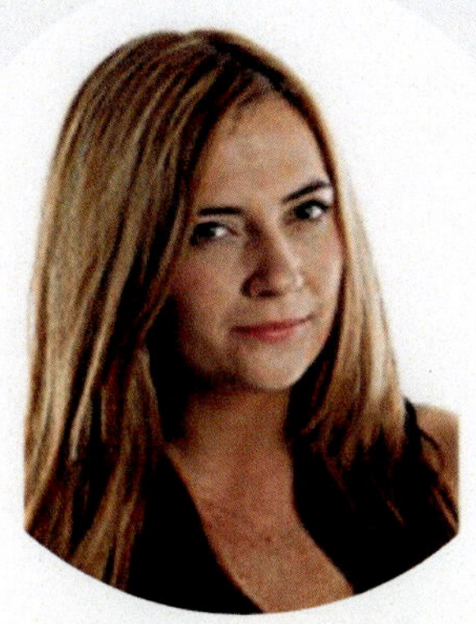

+ Follow

Follow to get new release updates and improved recommendations

About Lolly Brown

A life-long animal lover, Lolly Brown is equally comfortable writing about exotic creatures like the Mexican axolotl or dispensing practical advice to dog owners about kennel cough.

As a child, Brown first learned about fish and aquaria when her father brought home a 10-gallon aquarium as a surprise for his daughter. Within months, the father-daughter team graduated to a 120-gallon tank and were immersed in the intricacies of tank population management.

"We had that go-big-or-go-home mentality common to the hobby," Brown said. "Now I look back and think about what we did to Mama's living room! She was very patient with us."

Brown's fascination with animals continued in college, where she took numerous field biology and wildlife classes that allowed her to view the behavior of many species in their native habitats.

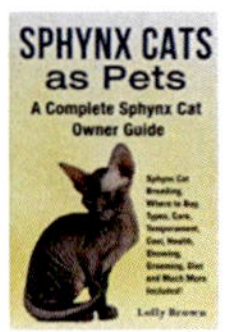

Paperback
$12.97

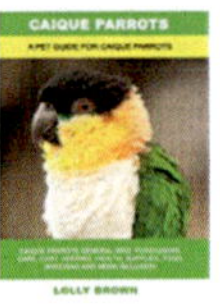

Paperback
$14.97

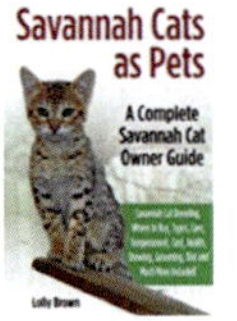

Paperback
$10.97

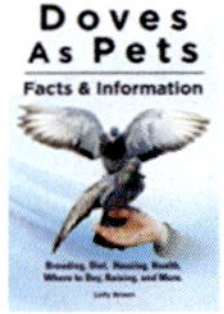

Paperback
$9.99

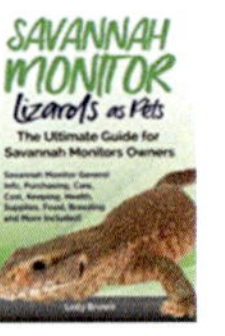

Paperback
$12.97

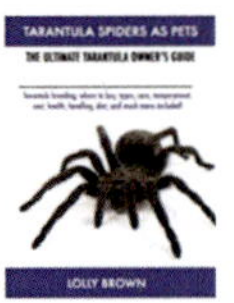

Kindle Edition
$3.49

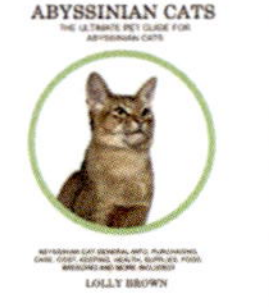

Kindle Edition
$3.99

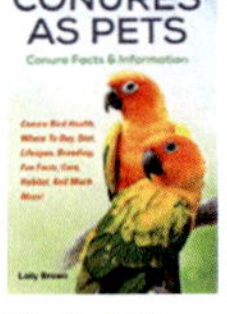

Kindle Edition
$6.97

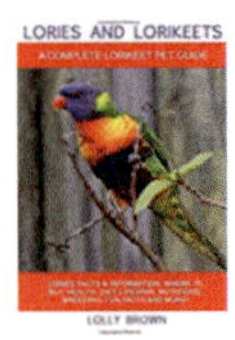

Paperback
$12.97

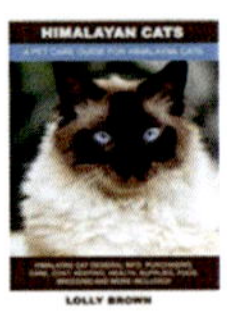

Paperback
$12.97

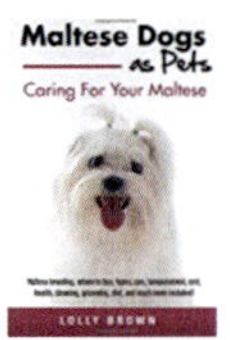

Paperback
$11.99

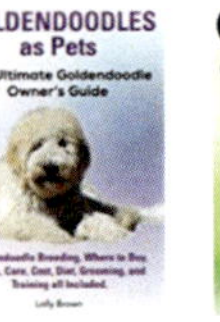

Paperback
$12.97

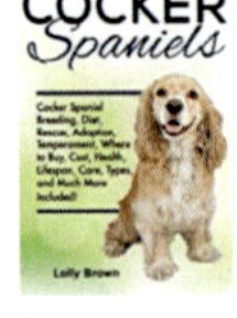

Paperback
$11.99

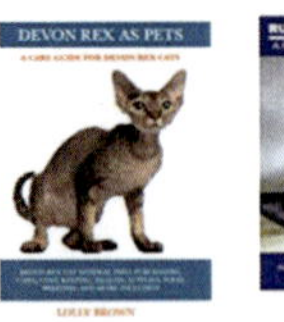

Kindle Edition
$4.49

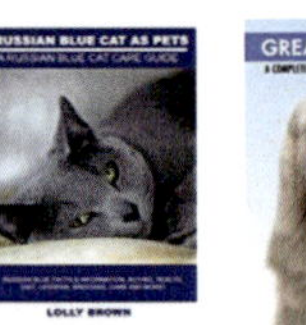

Paperback
$12.97

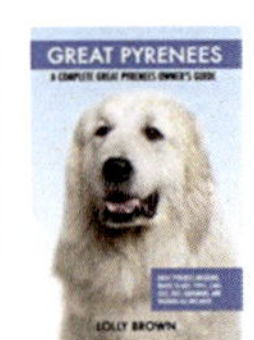

Kindle Edition
$3.99

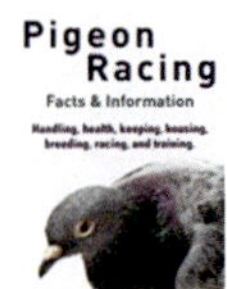

Paperback
$12.99

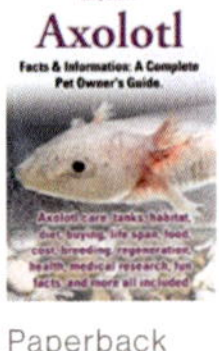

Paperback
$11.99

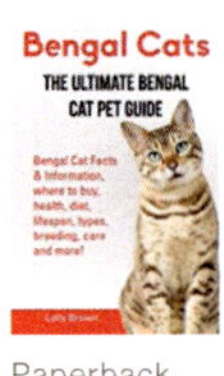

Paperback
$11.97

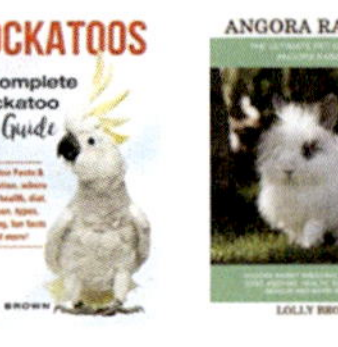

Paperback
$11.97

Kindle Edition
$3.99

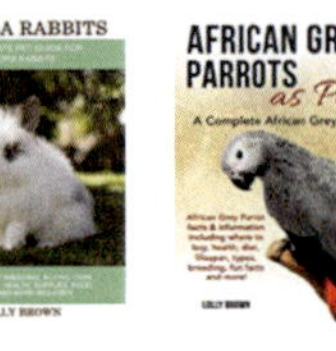

Paperback
$11.97

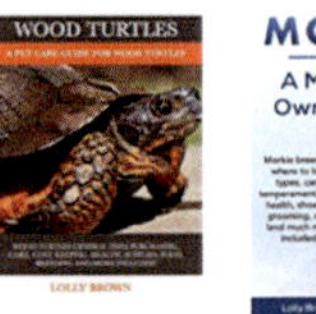

Kindle Edition
$3.99

Kindle Edition
$5.97

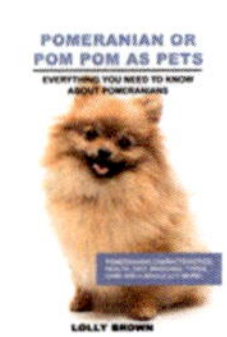

Kindle Edition
$0.00

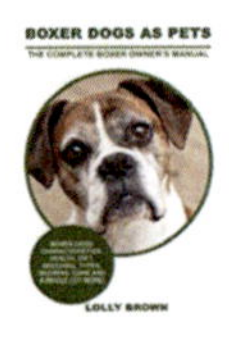

Kindle Edition
$3.99

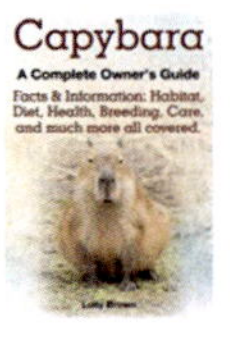

Paperback
$17.66

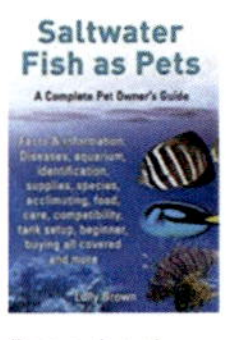

Paperback
$32.93

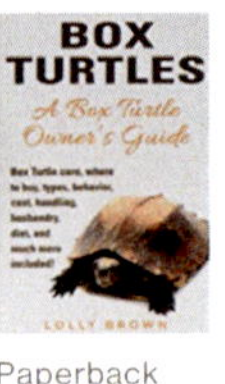

Paperback
$16.43

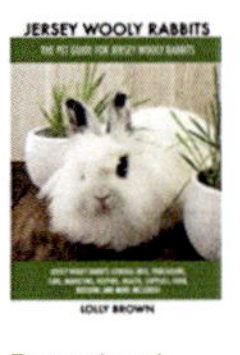

Paperback
$12.97

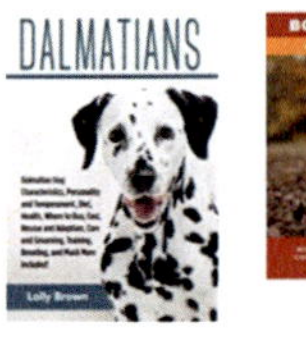

Paperback
$11.99

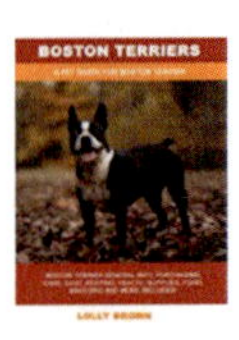

Kindle Edition
$3.99

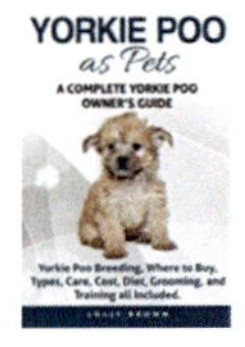

Paperback
$11.97

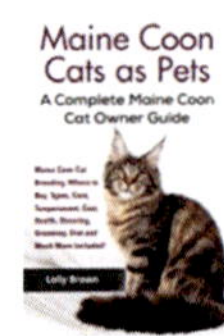

Paperback
$14.97

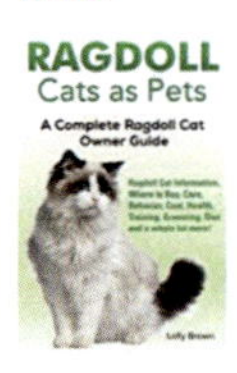

Paperback
$11.97

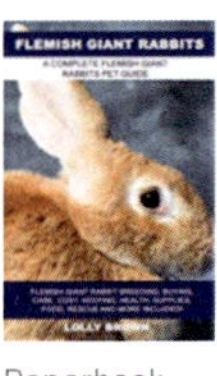

Paperback
$12.97

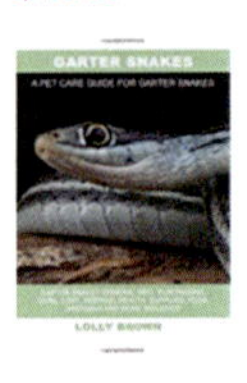

Paperback
$12.97

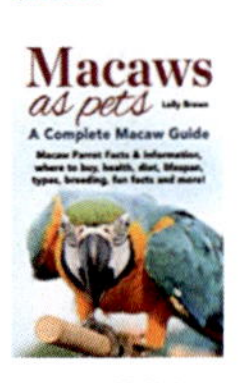

Kindle Edition
$3.99

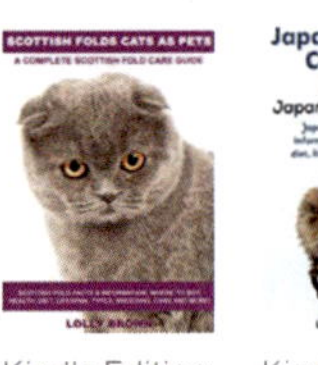

Kindle Edition
$3.99

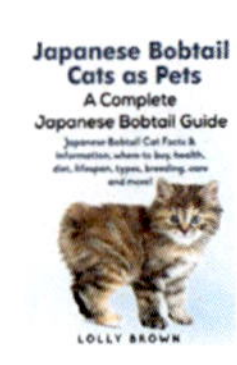

Kindle Edition
$3.49

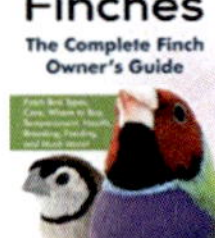

Paperback
$12.97

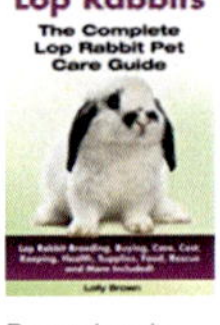

Paperback
$11.97

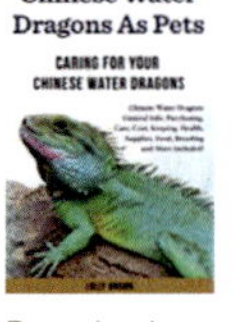

Paperback
$12.97

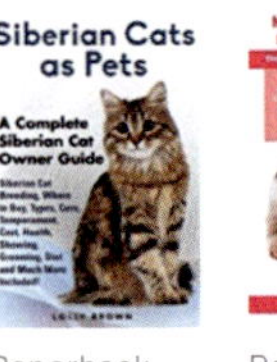

Paperback
$11.97

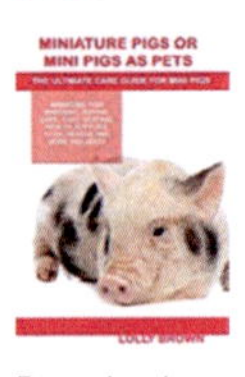

Paperback
$11.97

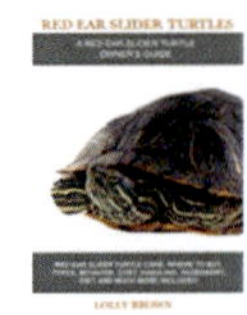

Paperback
$12.97

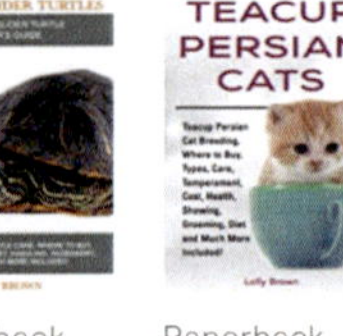

Paperback
$12.97

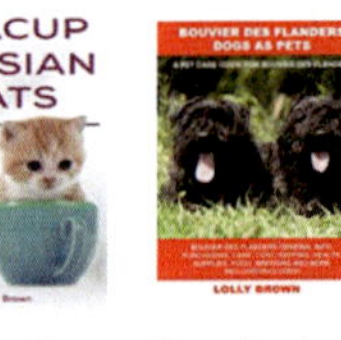

Paperback
$3.99

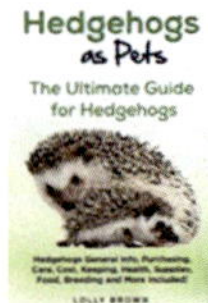

Paperback
$12.97

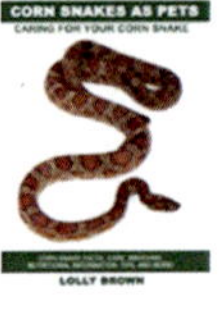

Paperback
$12.97

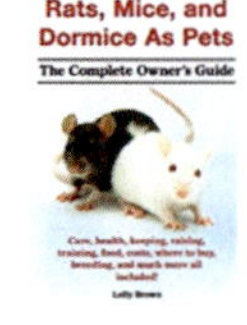

Paperback
$12.97

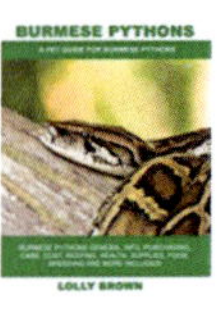

Kindle Edition
$4.97

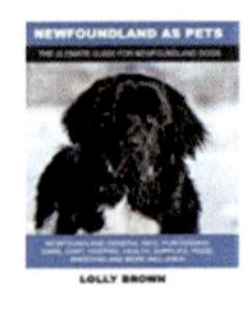

Paperback
$12.97

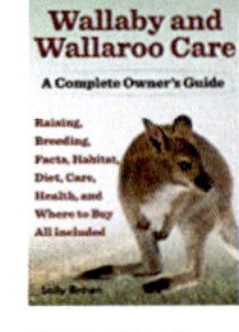

Paperback
$12.97

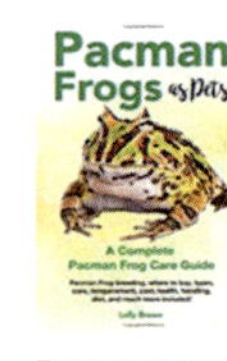

Paperback
$12.97

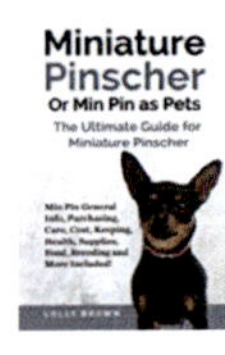

Paperback
$12.97

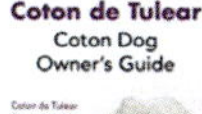

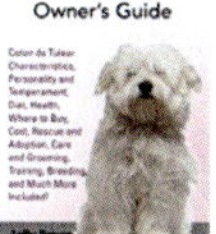

Kindle Edition
$3.99

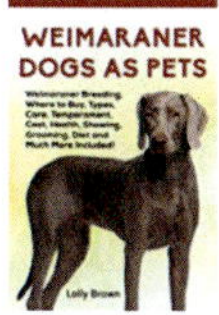

Kindle Edition
$3.49

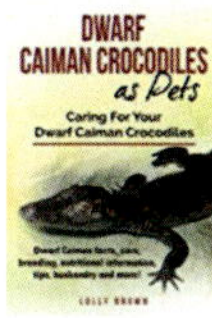

Paperback
$12.97

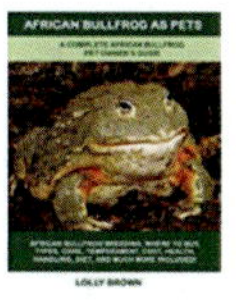

Paperback
$12.97

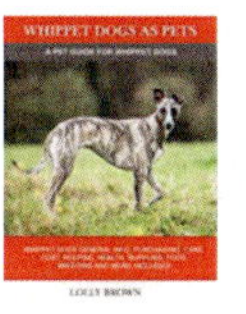

Kindle Edition
$4.99

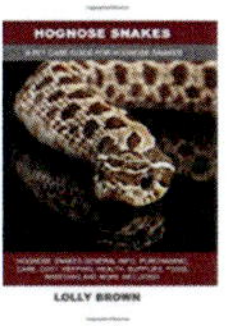

Paperback
$14.97

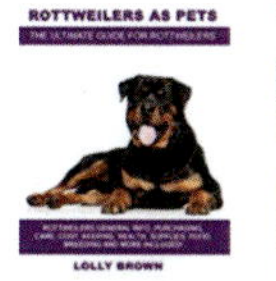

Kindle Edition
$4.49

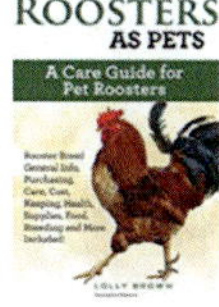

Paperback
$14.97

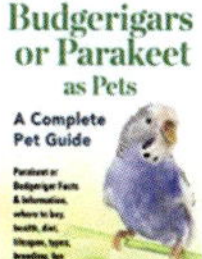

Kindle Edition
$3.49

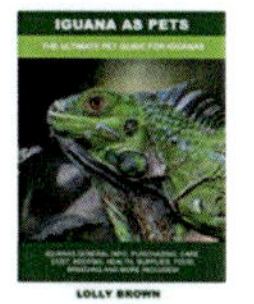

Paperback
$14.97

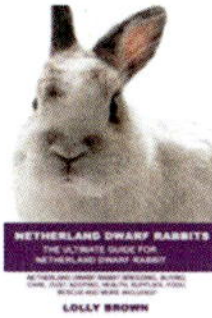

Kindle Edition
$3.49

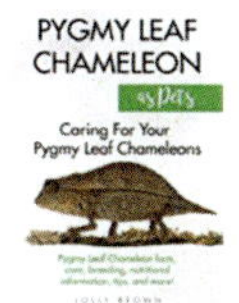

Kindle Edition
$3.49

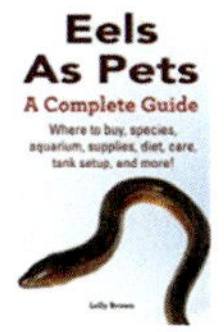

Paperback
$11.99

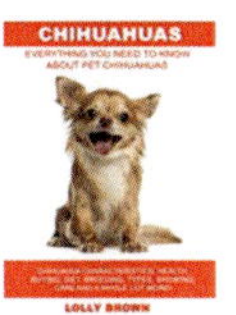

Kindle Edition
$3.49

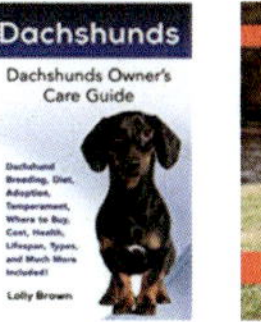

Kindle Edition
$3.49

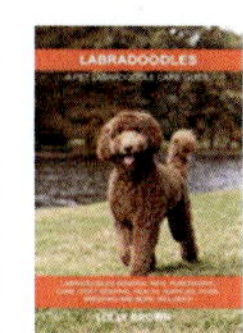

Kindle Edition
$4.99

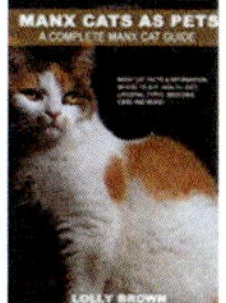

Paperback
$12.97

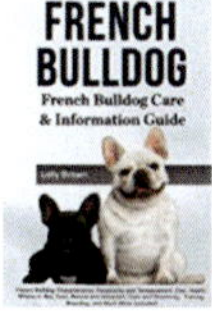

Kindle Edition
$3.49

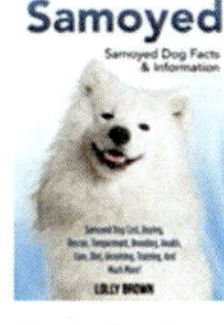

Paperback
$14.97

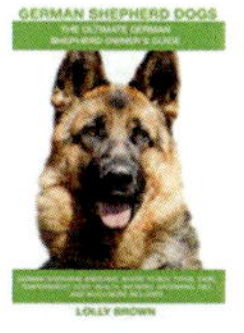

Kindle Edition
$3.49

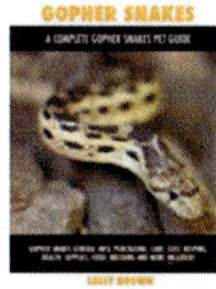

Kindle Edition
$3.99

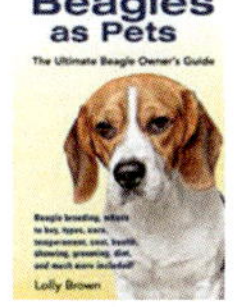

Kindle Edition
$3.49

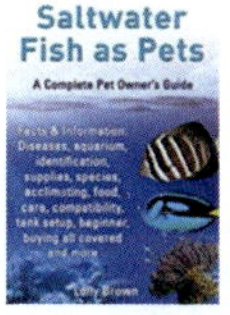

Paperback
$12.97

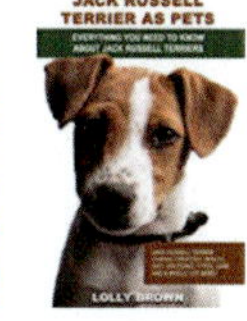

Kindle Edition
$3.49

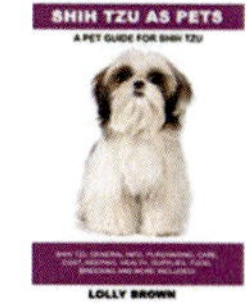

Kindle Edition
$4.97

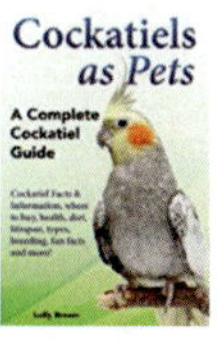

Kindle Edition
$3.49

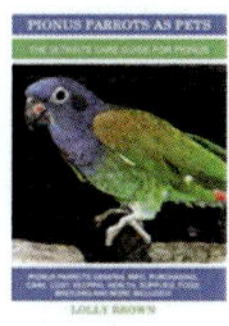

Kindle Edition
$4.97

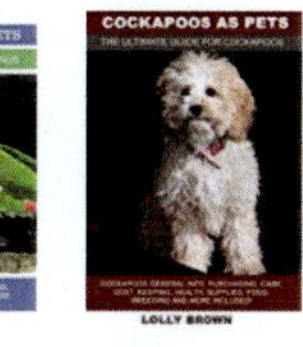

Kindle Edition
$3.99

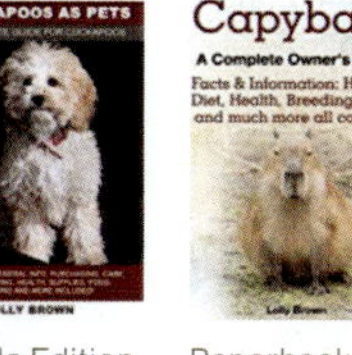

Paperback
$12.97

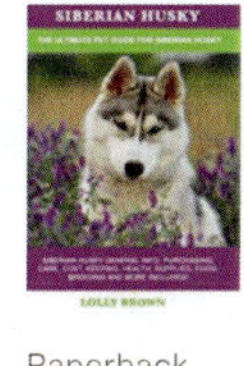

Paperback
$14.97

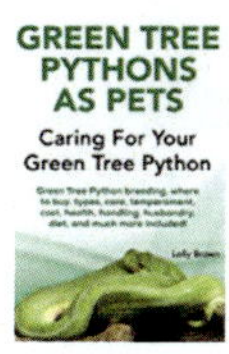

Kindle Edition
$3.49

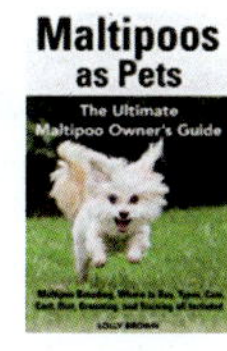

Kindle Edition
$3.99

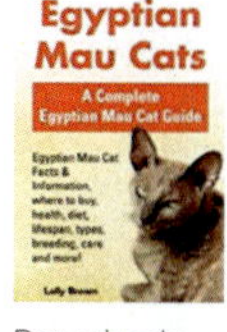

Paperback
$11.97

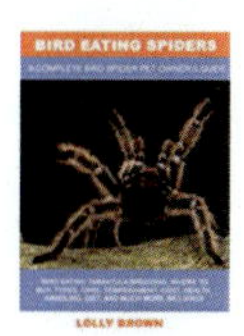

Kindle Edition
$3.99

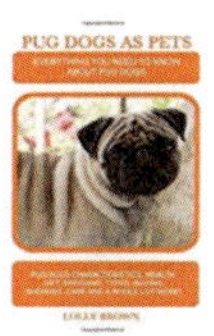

Paperback
$12.97

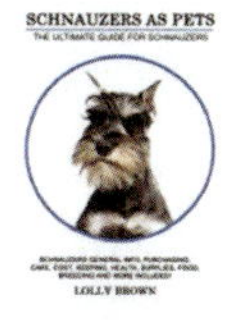

Kindle Edition
$3.99

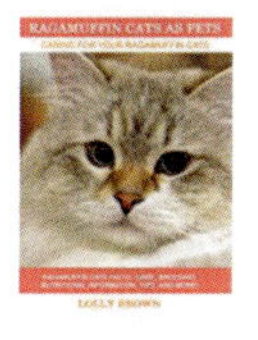

Kindle Edition
$4.49

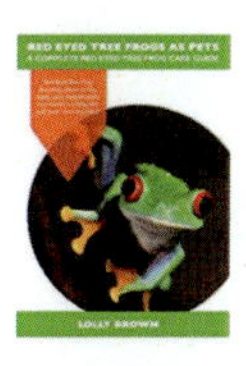

Kindle Edition
$3.49

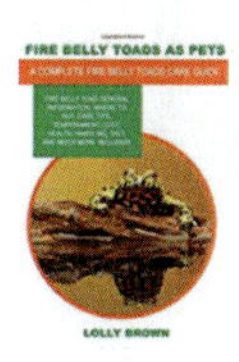

Paperback
$12.97

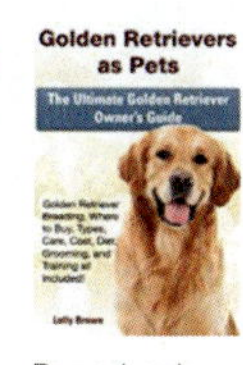

Paperback
$10.99

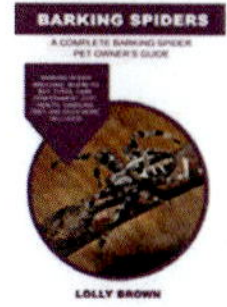

Kindle Edition
$3.49

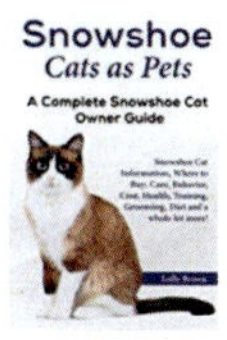

Kindle Edition
$0.00

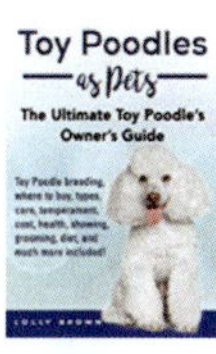

Paperback
$11.99

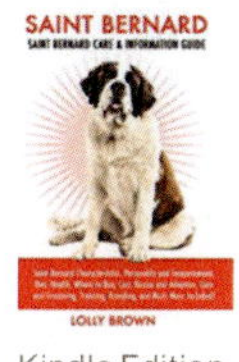

Kindle Edition
$3.49

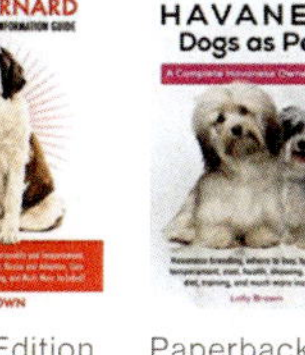

Paperback
$11.99

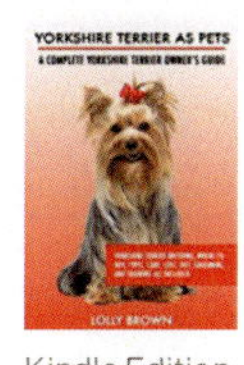

Kindle Edition
$3.49

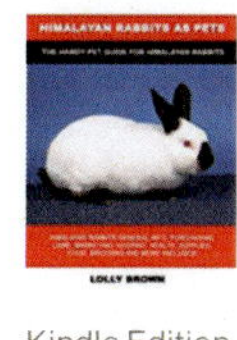

Kindle Edition
$3.99

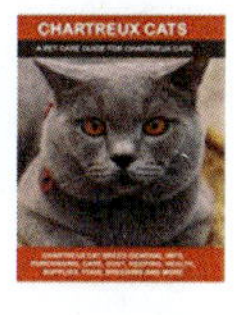

Kindle Edition
$4.99

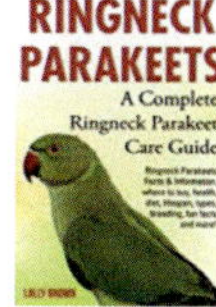

Paperback
$11.97

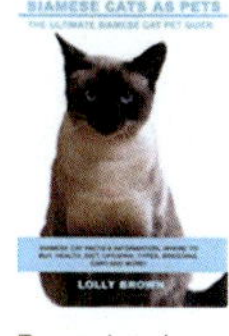

Paperback
$12.97

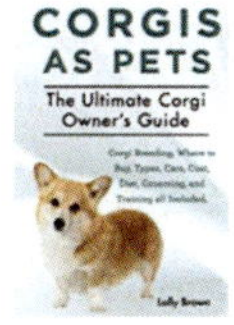

Paperback
$12.97

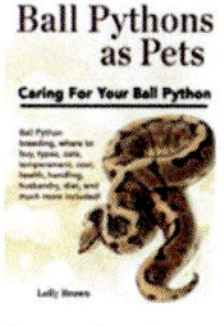

Paperback
$11.97

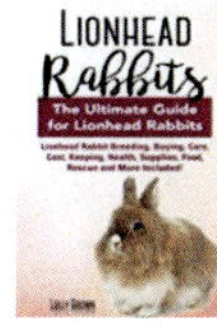

Paperback
$14.97

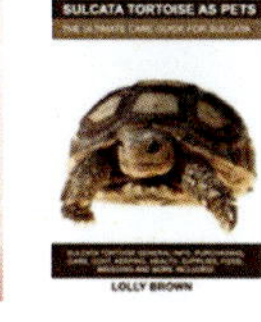

Paperback
$12.97

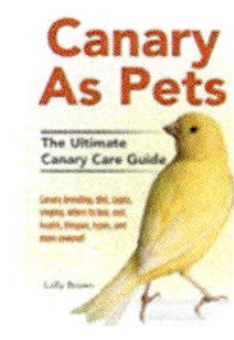

Paperback
$11.99

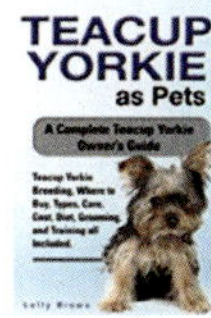

Paperback
$11.97

A Network of Pet Gurus

Authors with mirror identities

Two writers connected to Lolly Brown through Amazon.com publish books about more typical domestic pets. The books of these authors, Susanne Saben and Mark Manfield, have bizarrely similar covers. Their biographies are also the same except Manfield is from *central* Georgia and Susanne is from *southern* Georgia.

Most likely all of the books are written by ghostwriters like MS. It is possible that this network of pet manual writers leads back to the same person.

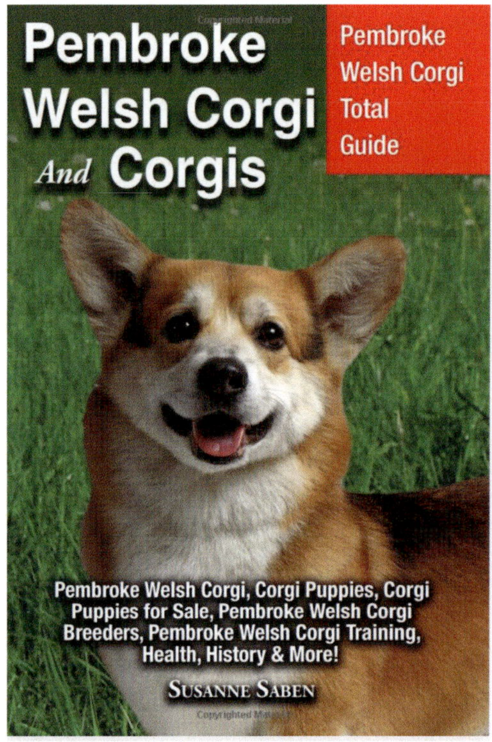

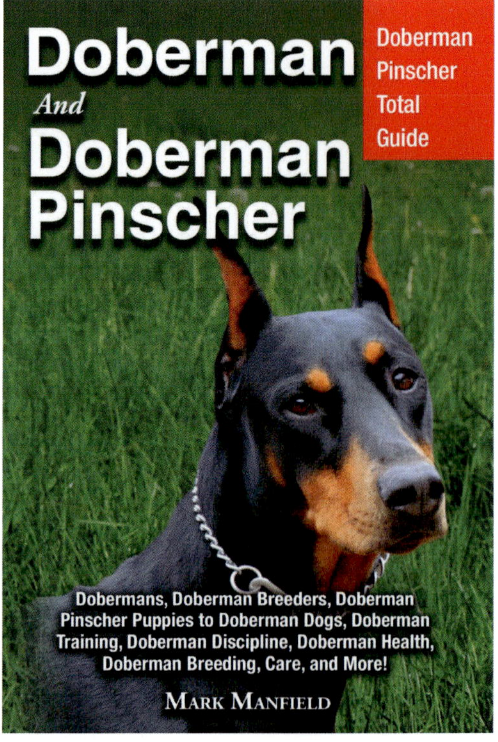

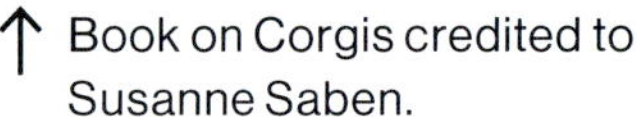
↑ Book on Corgis credited to Susanne Saben.

↑ Book on Doberman Pinschers credited to Mark Manfied.

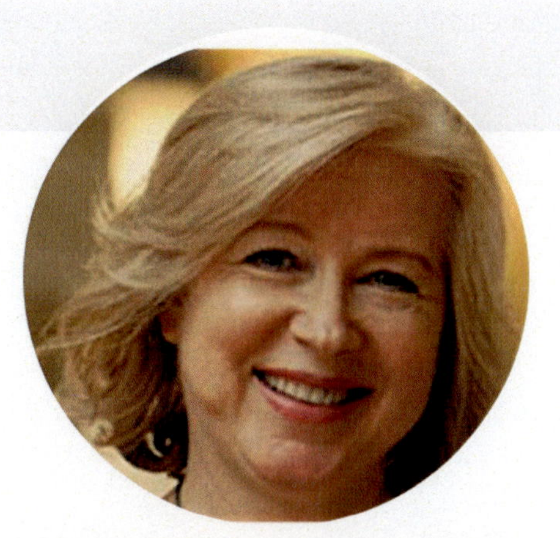

+ Follow

Follow to get new release updates and improved recommendations

About Susanne Saben

Susanne Saben is a dog lover and owner who has years of experience advising on all aspects of your dog's care from breeders, adoption and rescue, raising, feeding, care, temperament, health, and having the best possible experience with your dog!

Susanne writes in an entertaining and straightforward style which you will be sure to enjoy while you benefit from her years of experience and love for the animals she so richly writes about. Susanne lives in southern Georgia with her 3 children and husband, and has been a writer and avid animal lover for many years!

^ Read less

↑ Biography of Susanne Saben, pet manual author on Amazon.com.

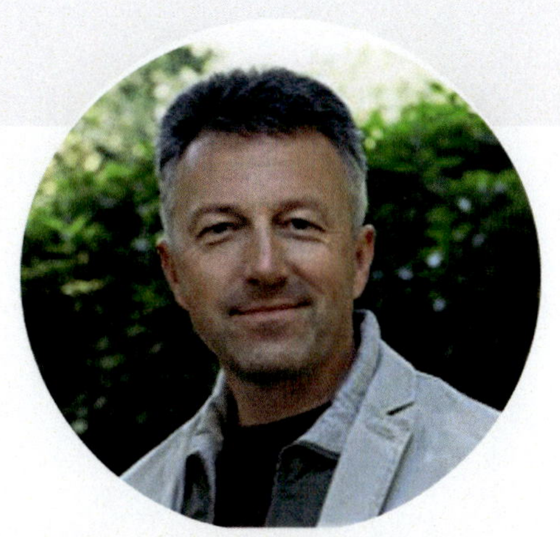

+ Follow

Follow to get new release updates and improved recommendations

About Mark Manfield

Mark Manfield is a dedicated dog lover and dog owner who has years of experience advising on all aspects of your dog's care from breeders, adoption and rescue, raising, feeding, care, temperament, health, and having the best possible experience with your dog.

Mark writes in an entertaining and straightforward style which you will be sure to enjoy while you benefit from his years of experience and love for the animals he writes about. Mark lives in central Georgia with his wife, and has been a writer and avid animal lover for many years.

^ Read less

↑ Biography of Mark Manfield, pet manual author on Amazon.com.

Anonymous Fans

An infinite circle of secret identities

To increase their book's Amazon.com rankings, self-published authors need reviews. Reviews can be written by paid ghostwriters or other self-publishers who want to exchange reviews for mutual benefit.

Paid reviews are an exercise in creative writing. Writers of paid reviews often describe their state of mind when they purchased the book rather than mentioning the content of the book itself. Other times they use generic descriptors like "informative" and "interesting" or how the book has "all the information you need." In stark contrast to these glowing appraisals are more believable reviews that describe the book's disappointing content, poor grammar, and typos.

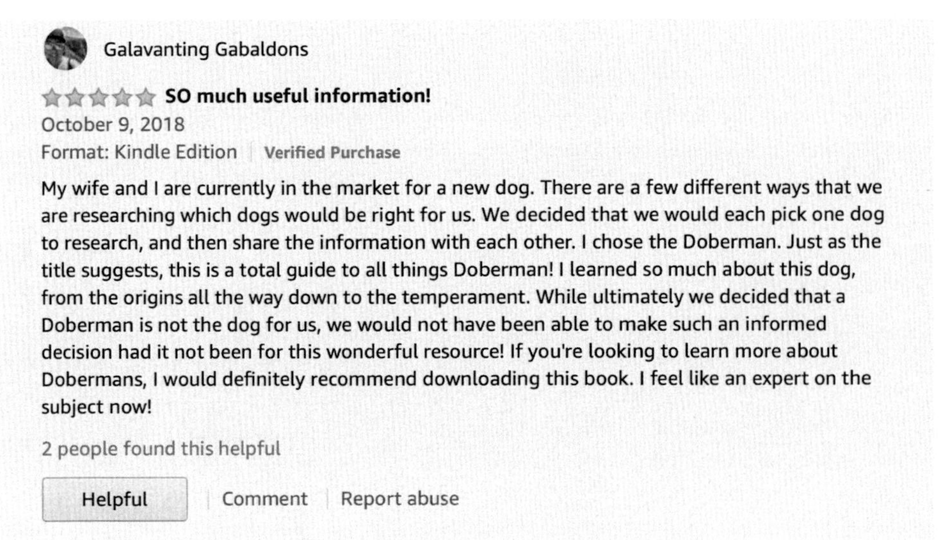

Galavanting Gabaldons

SO much useful information!

October 9, 2018

Format: Kindle Edition | Verified Purchase

My wife and I are currently in the market for a new dog. There are a few different ways that we are researching which dogs would be right for us. We decided that we would each pick one dog to research, and then share the information with each other. I chose the Doberman. Just as the title suggests, this is a total guide to all things Doberman! I learned so much about this dog, from the origins all the way down to the temperament. While ultimately we decided that a Doberman is not the dog for us, we would not have been able to make such an informed decision had it not been for this wonderful resource! If you're looking to learn more about Dobermans, I would definitely recommend downloading this book. I feel like an expert on the subject now!

2 people found this helpful

Helpful | Comment | Report abuse

↑ Creative, positive customer review for *Doberman and Doberman Pinscher Total Guide* by Mark Manfield (pictured on p.24).

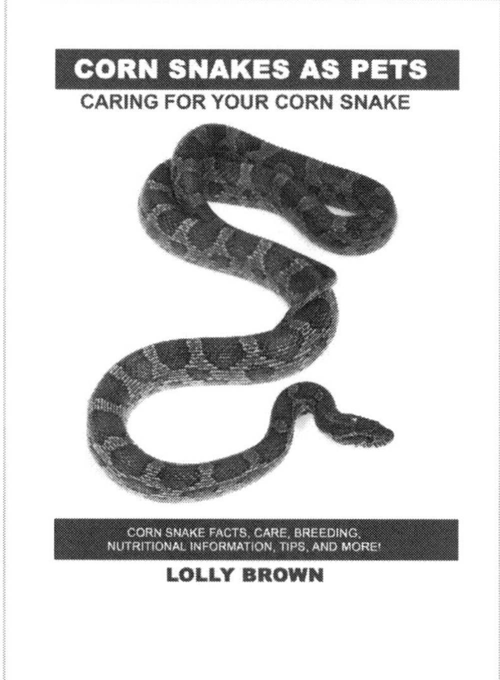

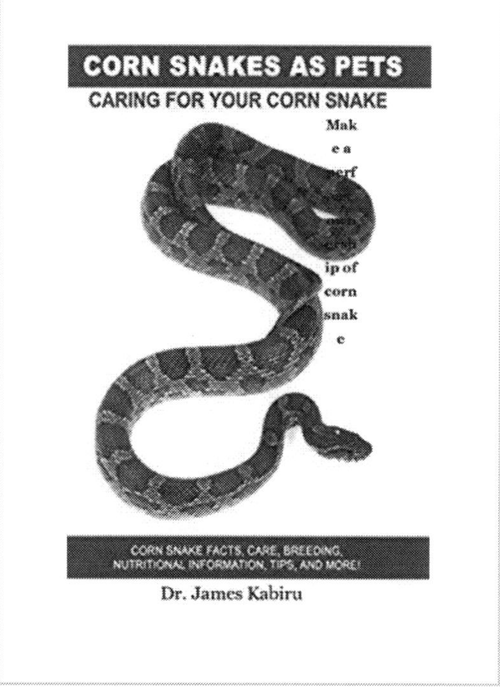

↑ Identical books about pet corn snakes published under different author names: Lolly Brown and Dr. James Kabiru.

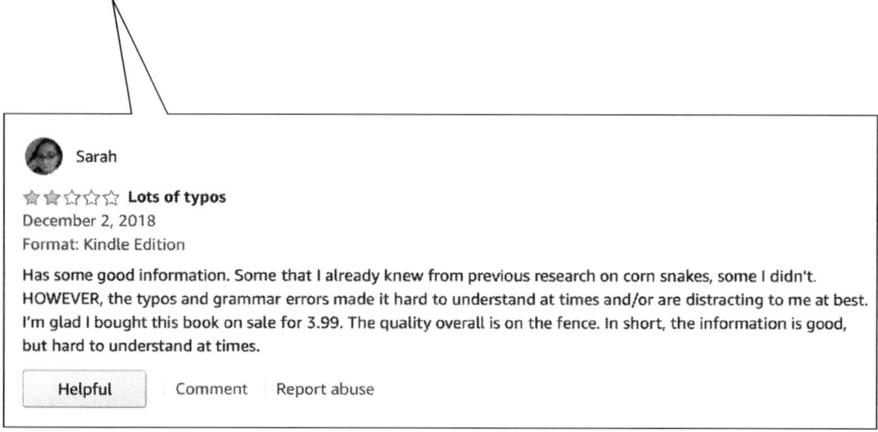

Sarah

Lots of typos

December 2, 2018

Format: Kindle Edition

Has some good information. Some that I already knew from previous research on corn snakes, some I didn't. HOWEVER, the typos and grammar errors made it hard to understand at times and/or are distracting to me at best. I'm glad I bought this book on sale for 3.99. The quality overall is on the fence. In short, the information is good, but hard to understand at times.

Helpful | Comment | Report abuse

↑ Negative customer review for *Corn Snakes as Pets: Caring for Your Corn Snake* by Lolly Brown.

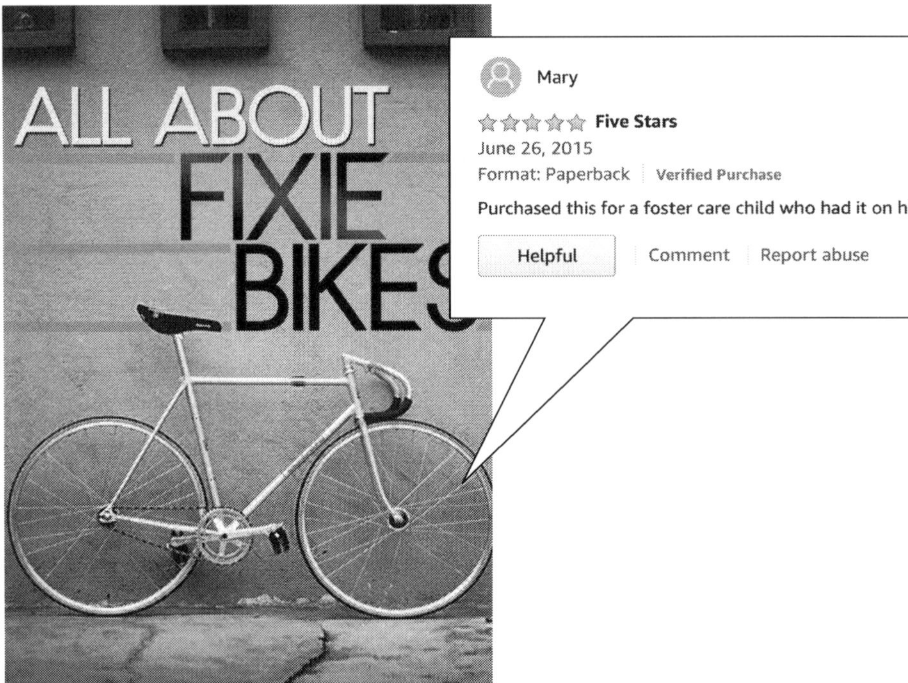

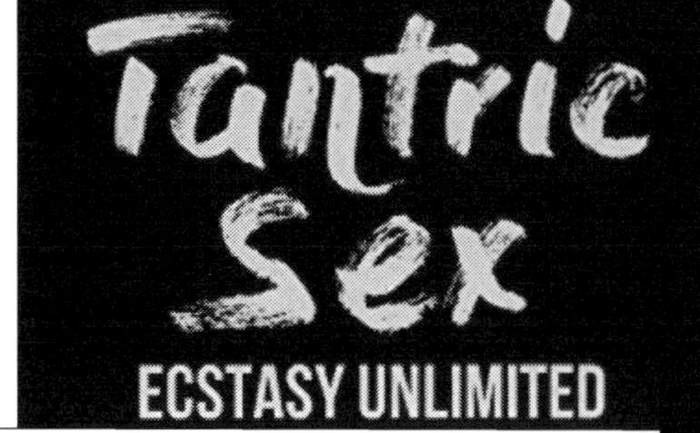

 George

Reader can easily understand and follow the sex positions

April 3, 2016

Verified Purchase

I had bit knowledge about tantra and tantric sex. I purchased this book for knowing more sexual position that will be helpful for me and my partner. This book is a big collection of sexual positions according to Kama Sutra. I am impressed with all the positions that make my partner jollier. Reader can easily understand and follow the sex positions. I think it is indispensable in your bedroom. Thanks Sophy.

Helpful | Comment | Report abuse

 Selection of ghostwritten books and their reviews on Amazon.com.

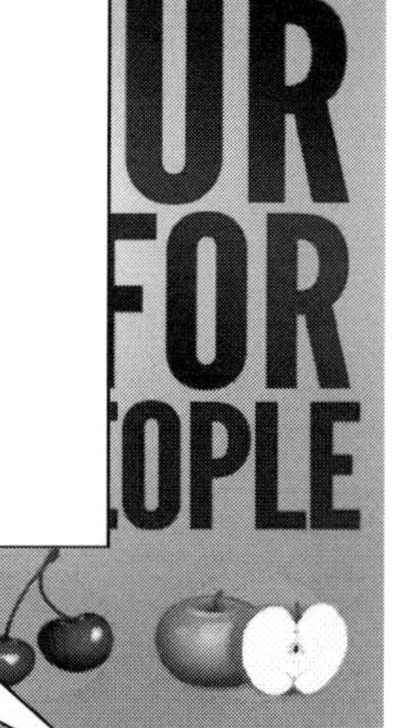

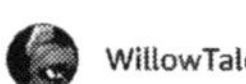

WillowTale

Good book but...

December 2, 2015

Format: Kindle Edition | Verified Purchase

This is a quick read with valuable information. The 8 hour diet does work. So why 4 instead of 5 stars? I find it very irritating to read anything that is so much in need of proofreading and copy editing. This is one of those books. One or two mistakes I can understand. Multiple errors throughout the book, not so understanding. And sorry, a low price is no excuse for sloppy writing. I would ask the author to please find someone who will actually read and correct this for grammar (or carefully re-do it himself). Then it will be worth 5 stars.

3 people found this helpful

Helpful | Comment | Report abuse

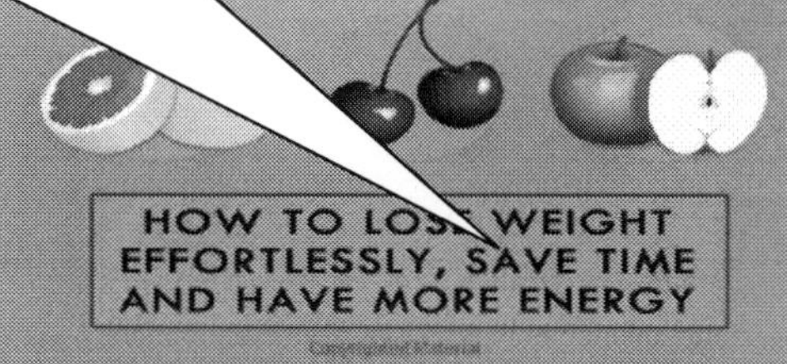

Attract

MONEY FAST

Suzanne Gonzalez

The easiest guide.

June 17, 2019

Format: Kindle Edition | Verified Purchase

There are numerous connections in the book with helpful data and locales that the writer utilizes himself to invest.This is a relative snappy perused with little fluff.If more top to bottom data is required the writer gives his contact data and a rundown of different books that go into more subtleties of the systems gave in this book.Overall a decent spot to start.The Author well composes it.I am not looking at turning into an informal investor watching a screen throughout the day or somebody affixed to the financial exchange scroll yet simply being somebody that knows about how patterns influence organizations and when to move or hold your position dependent on those. I picked this ideal manual for all.

Helpful | Comment | Report abuse

Profits and Losses

Amazon's shifting marketplace

In the conclusion of his Skillshare.com class, NK encouraged everyone to follow his blog and see his progress. In a 2015 blog post, he set a goal to make the average monthly wage in Denmark ($3,399.48 a month) through book sales.

Around that time, Amazon changed the marketplace by introducing a new subscription model called Kindle Unlimited. This new service allows Amazon customers to borrow any number of Kindle eBooks for a monthly subscription fee. Kindle authors are given a flat fee for these loans. While they most likely see more downloads of their books than before, their profits have declined dramatically.

Let me explain **$4,260** x 0.6 (US store 60% of income) = **$2,556**. **$2,556** x 1.33 (I need the results for four weeks, not six) = **$3.399.48.**

So yesterday when my sales were updated the result was **$1,534.60** for the US store.

Avg. List Price($)	Avg. File Size (MB)	Avg. Offer Price($)	Avg. Delivery Cost($)	Royalty (USD)
				Total: $ 1,534.60
4.99	0.27	6.66	0.00	10.50
2.99	0.51	2.99	0.08	2.04
4.03	0.42	4.03	0.06	25.03

So I need this number to be **$3.399.48** by September. So with **$1,534.60** I am **45.14%** of reaching my goal.

With 26 weeks to go I am in a hurry. I just commissioned 4 books over the last couple of days. Trying my luck with children's books this time. Excited to see the results.

So that's that for this week. I will return next week with an update:)

Don't be a stranger. Thoughts, input and opinions much appreciated in the comments below!

It's a Trap! (How Kindle Unlimited is Hurting Authors... While Kindle Select Prevents Them From Leaving)

by R.M. ArceJaeger | Feb 4, 2015 | Publishing

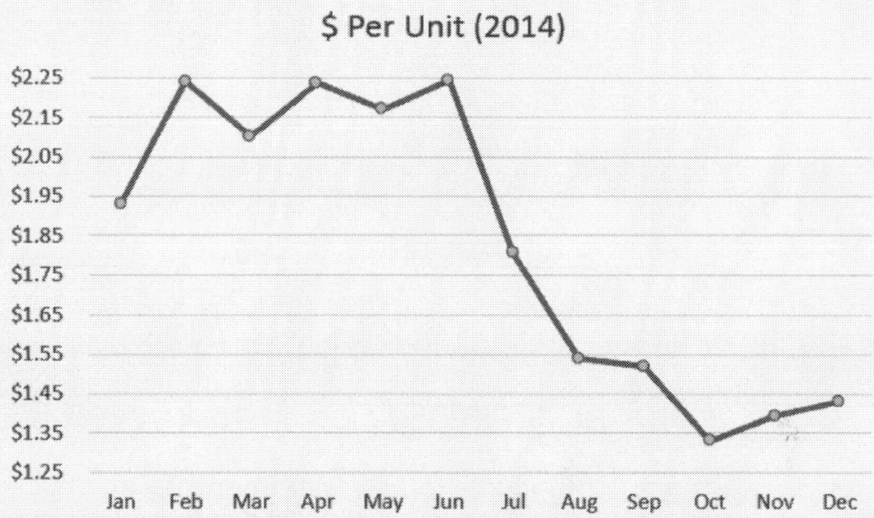

When Amazon first rolled out its author payment system for Kindle Unlimited, I was ecstatic. Kindle Unlimited lets readers borrow books for free, sort of like a digital library. As an author, any book I had signed up through Kindle Select (a special program for self-publishers) would automatically be made available for borrows through Kindle Unlimited. True, I would not be receiving any sales profit from people who read my books via KU, but that was alright because Amazon was paying me for each borrow instead—and what they were paying per book was actually *more* than the royalties I would have received from a sale.

At least, it *was*. I've been tracking all my sales data ever since 2012, and I've noticed a definite trend. KU sales are going up (to the point where they outnumber my "real" sales by a factor of six) whereas Amazon's payments for borrows are going down . . . way down.

↑ Blog post from Kindle author R.M. ArceJaeger explaining how the Kindle Unlimited service caused a steep decline in book revenue.

← Screen capture from NK's blog showing his sales figures. In an effort to reach his goal, he commissioned four books two days prior to his post. He also started experimenting with children's books.

Jokes and Gags

Amazon's rules for content

In addition to changes in the Kindle payment structure, Amazon started enforcing rules for what constitutes a "quality" book. To meet Amazon's content standards, the book has to have a minimum word count of 2,500 words.* Even books that have already been published could be removed from the website if they don't meet this criteria. In the words of Amazon, they "create a poor reading experience."

For self-published paperback books on Amazon.com, there are ways around the word count requirements. A book can have repeated content or blank pages so long as each page has a header and the product description clearly labels the book as a "joke" or "gag book."

*A series of books I made with computer-generated text did not originally meet the 2,500 word count threshold. To reach Amazon's quality standards, I changed a line in my code to loop more times, and in turn, output more text.

Print Publishing Guidelines

kindle direct publishing

Genre-specific Requirements

Joke books

Joke or gag books with repeated content or an intentional absence of content can be published as long as they are clearly labeled as such in the product description and they meet all other specification requirements. Books meant to contain empty pages should include some type of content such as lines, headers, or "notes" to indicate the pages are intended to be blank.

↑ Amazon's guidelines for "joke books."

Hello,

During a quality assurance review of your KDP catalog we have found that the following book(s) are extremely short and may create a poor reading experience and do not meet our content quality expectations:

Name of Short

In the best interest of Kindle customers, we remove titles from sale that may create a poor customer experience. Content that is less than 2,500 words is often disappointing to our customers and does not provide an enjoyable reading experience.

We ask that you fix the above book(s), as well as all of your catalog's affected books, with additional content that is both unique and related to your book. Once you have ensured your book(s) would create a good customer experience, re-submit them for publishing within 5 business days. If your books have not been corrected by that time, they will be removed from sale in the Kindle Store. If the updates require more time, please unpublish your book.

↑ *Topic: Amazon going after short shorts*, Kboards, https://www.kboards.com/index.php/topic,149460.0.html.

Books as Money

Amazon seems like Sotheby's

After my publishing bonanza of ghostwritten and computer-generated books had passed, I almost forgot these projects existed on Amazon.com. Notifications of deposits for purchases of the books occasionally popped up in my inbox, but my profits were just a few dollars here and there. None of my books were providing the passive income that NK, the Danish male model, had bragged about.

A few years after I published *What Does the Bible Say About Word Finds?*, I looked it up on Amazon.com to share with a friend. I was surprised to find that GoldieLoxBooks, an Amazon reseller, was selling a used copy of my book for $2,796 even though new copies were still available for $10.

For a brief moment I thought my book's identity as an art object had been recognized by GoldieLoxBooks. This was a passing thought and I settled on the most likely explanation: that the price was a computer glitch.

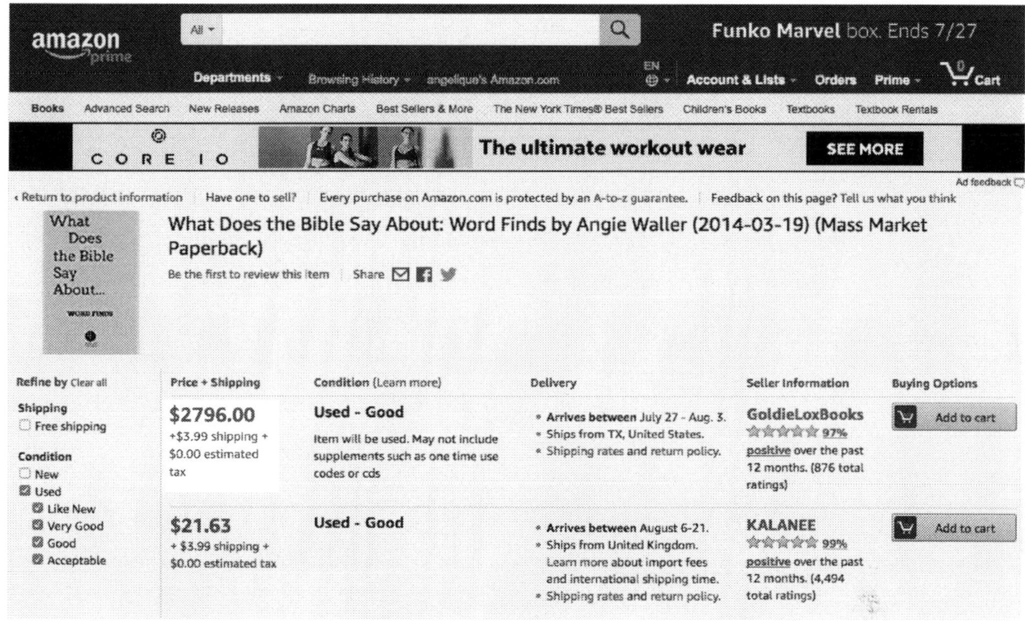

↑ My book *What Does the Bible Say...* listed on Amazon.com for $2,796 while still available new for $10.

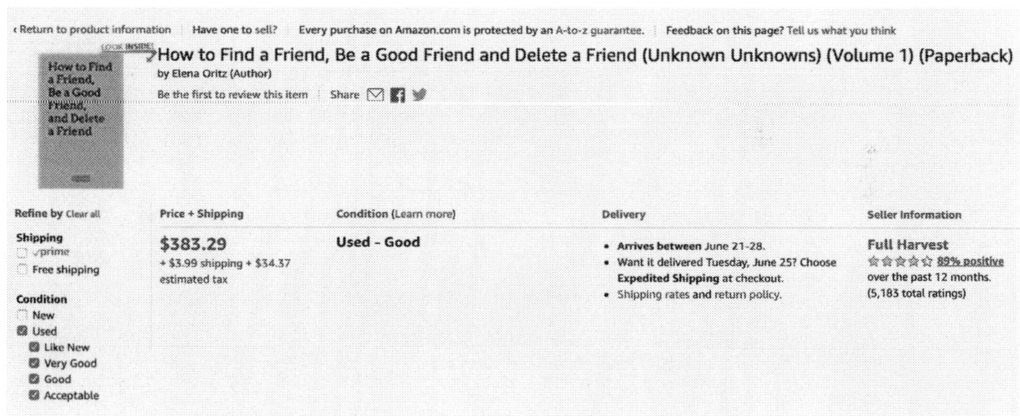

↑ My book *How to Find a Friend...* listed on Amazon.com for $383.29 while still available new for $10.

Marked Up Prices Are Trending

Books have a life of their own

More than a year after noticing my book's mysteriously high price on Amazon.com, stories about other people's books getting marked up were in the news. Well-known authors were quoted in the *New York Times* presuming high prices for their books were a result of their fame. Books being limited or early editions was also an assumed logic. Curiously, no one actually selling used paperbacks for thousands of dollars was interviewed for the article. Instead, authors and marketing experts shrugged it off as a by-product of capitalism.

In April 2018, three months prior to the *Times* article, *KrebsOnSecurity* (krebsonsecurity.com) and *The Guardian* also wrote about books on Amazon.com being sold at inflated prices. In this case, the 10-fold price markup was not on used books from well-known authors or publishing houses; rather, the original price set by the publisher on CreateSpace was $555.

Like other titles from CreateSpace, the $555 book was written under a pseudonym. According to *KrebsOnSecurity*, the book was filled with "gibberish," most likely computer-generated. The book in question, *Lower Days Ahead*, was not intended for reading.

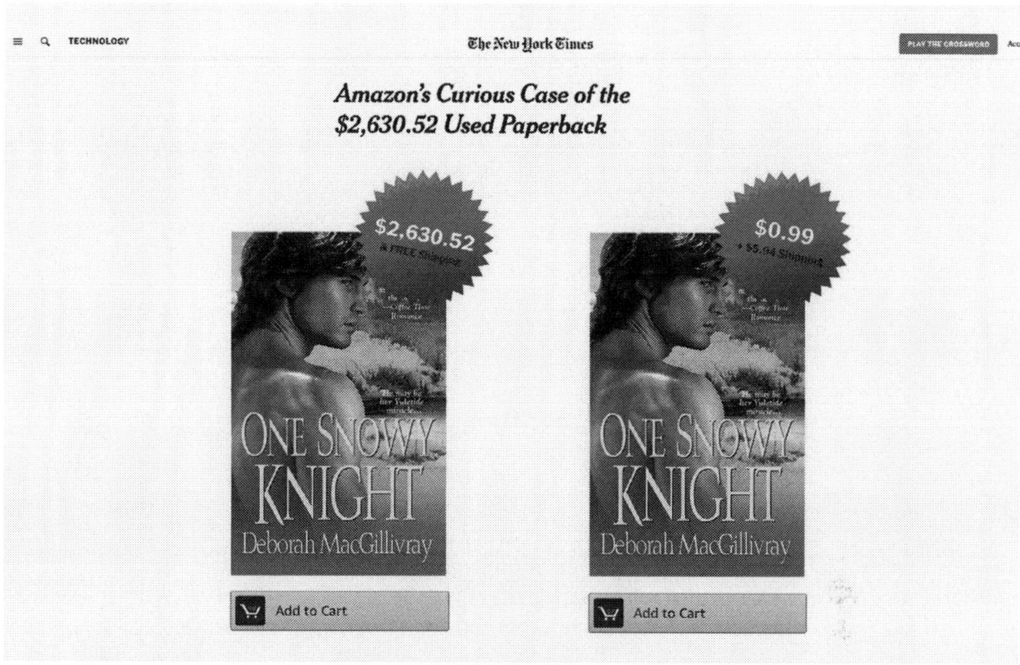

↑ "Amazon's Curious Case of the $2,630.52 Used Paperback," *The New York Times,* David Streitfeld, July 15, 2018.

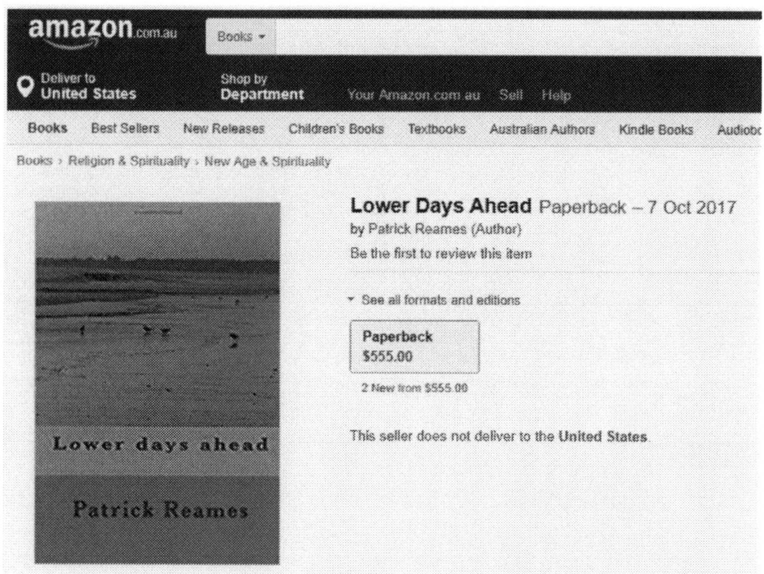

↑ *Lower Days Ahead*, a CreateSpace publication listed on Amazon.com for $555. (screen capture from krebsonsecurity.com)

Ghost in the Shell Game

Books are being traded for cocaine by drug cartels (possibly)

CreateSpace's lack of rules for author's name and publishing company allowed fictional personae like Lolly Brown to build pet book empires without readers knowing whether or not there is a real Lolly Brown. In the case of *Lower Days Ahead* listed for $555, Patrick Reames credited as the author is a real person. But Reames is not the person who published the book on CreateSpace.

The seller of *Lower Days Ahead* set up a CreateSpace account with Reames's social security number. When books with Reames's name and tax id were sold, the taxable income was attributed to Reames but none of the actual money went to him. Until he received a 1099 form from the IRS for the $24,000 worth of books he allegedly sold, Reames was totally unaware of *Lower Days Ahead*.

Reames had limited success with Amazon's fraud support. At the time *KrebsOnSecurity* was covering the incident, the case was far from being resolved. Reaching Amazon support agents on the phone also proved to be difficult. Ironically, a self-publisher on Amazon.com claims to know the secret to getting through to Amazon's phone support and has published eBooks and audiobooks with bootleg 1-800 numbers.

Fake books sold on Amazon could be used for money laundering

Books of gibberish are listed on Amazon.com for thousands of dollars, with one author claiming his name was used to send almost $24,000 to a fraudulent seller

Alison Flood
Fri 27 Apr 2018 10.00 EDT

↑ "Fake Books Sold on Amazon Could Be Used for Money Laundering," *The Guardian,* Allison Flood, April 27, 2018.

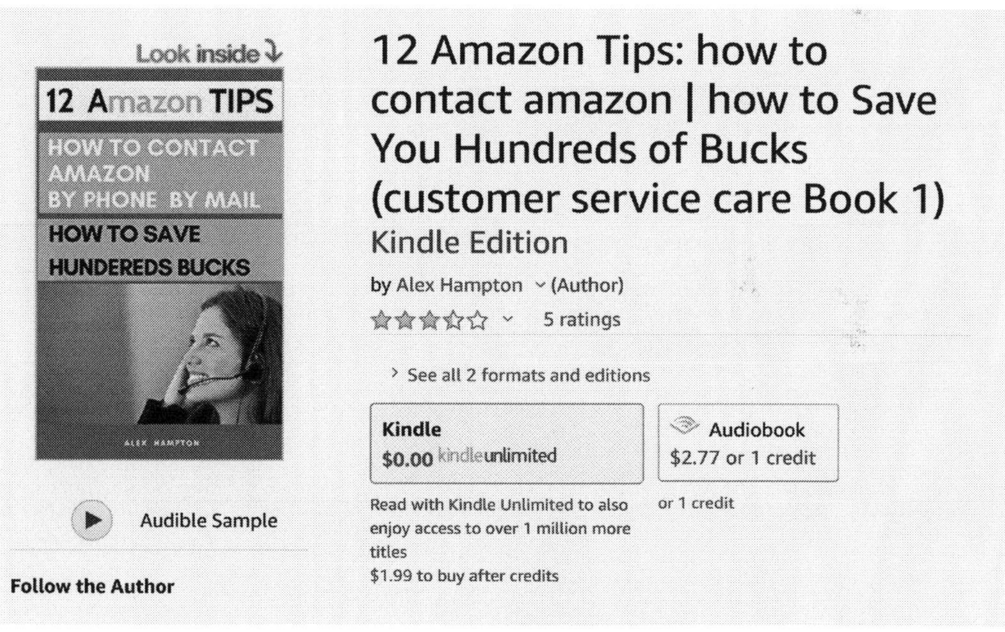

↑ Screen capture of self-published Kindle eBook and audiobook on Amazon.com claiming to have customer support numbers for Amazon.

Relentless

Books as decoys

My books are still on Amazon.com fluctuating from their original listing price to a few hundred and sometimes thousands of dollars. It is unlikely my book *What Does the Bible Say About Word Finds?* will sell for $1,000 unless that transaction is a stand-in for something else. While I cannot prove fraudulent activity is behind the prices of my books sold by resellers, it seems that laundering money through obscure used books would be an easy thing to do. No one will "accidentally" pay thousands of dollars for my book, but the seller can direct someone to "buy" it and earn their 60% commission on the sale with no questions asked. Even anonymous currencies like Bitcoin with weekly withdrawal limits present more red tape than this scenario.

For the sellers who list my books at high prices, I am unable to learn more about them. Like the books covered so far, these virtual storefronts have mixed reviews that are either extremely positive or complaining that an order was never received. While ordering my own book for $2,796 might lead to a postmarked envelope with a clue to the seller's location, there is no guarantee a package would arrive.

When Jeff Bezos conceived Amazon.com in 1994, his purpose for selling books online was to collect information about affluent book lovers who could later be customers for what the site would become.[1] Books were a prop to lure people in with their credit cards.

Most of the self-publishers I've described in this book are in tune with the Amazon ethos. For them, books are items to be produced at lightning speed and packaged to be impulsively purchased by unknown masses on the internet.

Although his friends discouraged him from calling his web store "Relentless," Bezos still registered the domain name relentless.com in 1994. If you type *relentless.com* or *amazon.com* in your web browser, you end up in the same marketplace of deception and secret identities I have described. You can also buy the latest *New York Times® Best Seller* and a subscription for organic diaper cream while you are there.

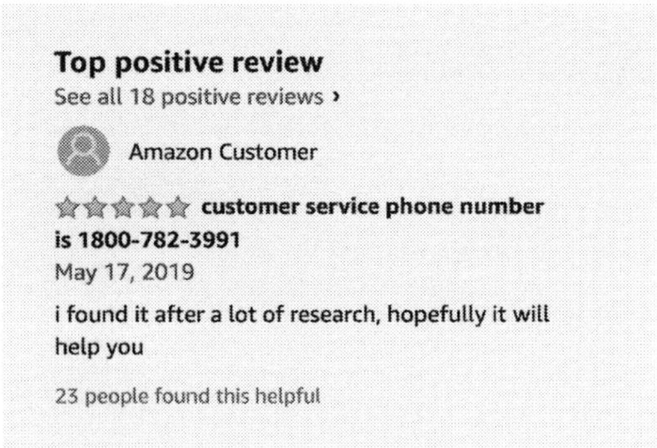

Top positive review
See all 18 positive reviews ›
Amazon Customer
customer service phone number is 1800-782-3991
May 17, 2019
i found it after a lot of research, hopefully it will help you
23 people found this helpful

↑ Screen capture of review listing questionable Amazon.com support number. As of Sept. 2019, a Google search for the number leads to Victorian Era & Country Creation in Forsyth, Montana.

1 George Packer, "Cheap Words: Amazon is good for customers. But is it good for books?", *The New Yorker*, Feb 9, 2014, *https://www.newyorker.com/magazine/2014/02/17/cheap-words.*

2 Brad Stone, *The Everything Store: Jeff Bezos and the Age of Amazon*, New York, October 15, 2013.

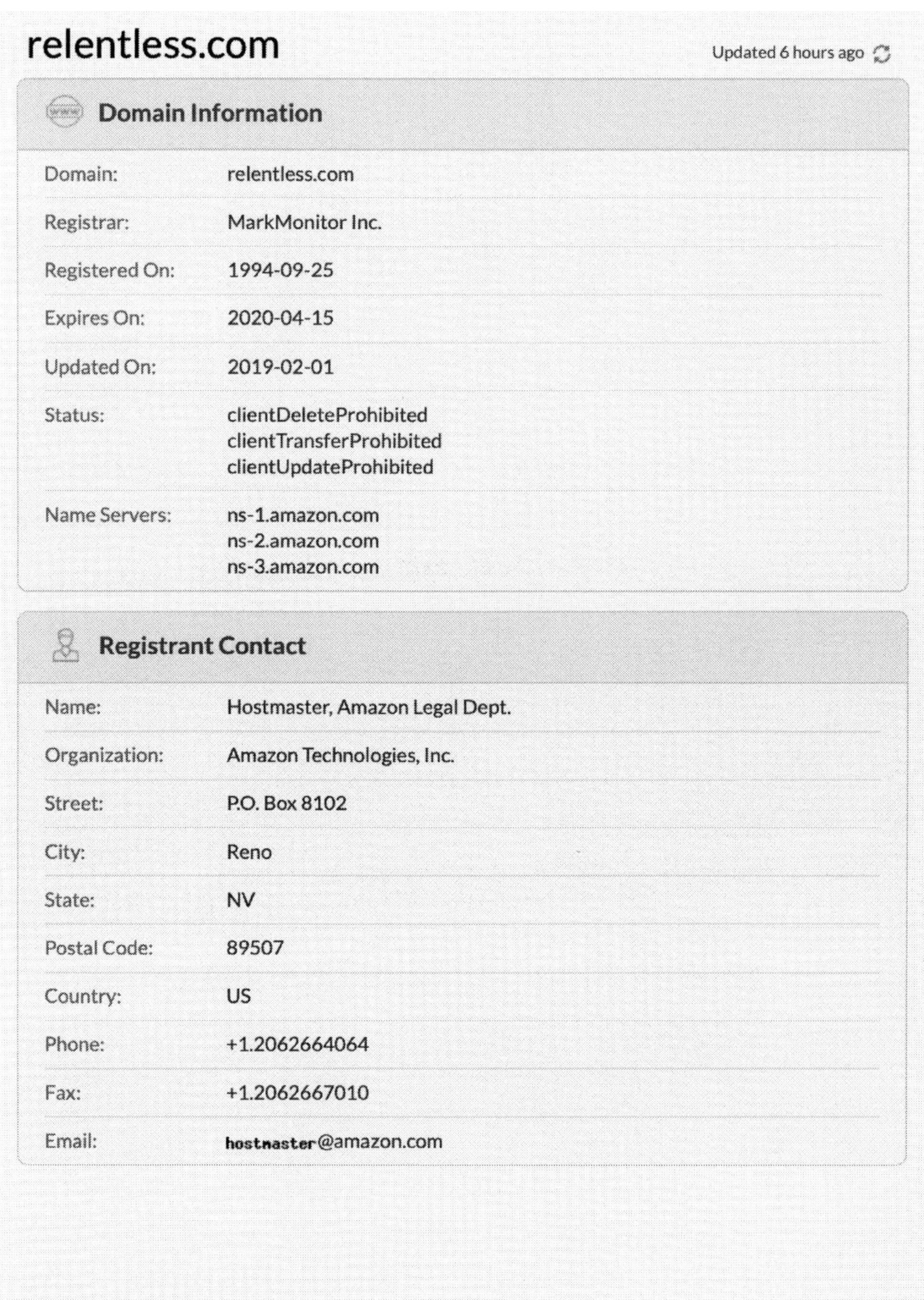

relentless.com

Updated 6 hours ago

Domain Information

Domain:	relentless.com
Registrar:	MarkMonitor Inc.
Registered On:	1994-09-25
Expires On:	2020-04-15
Updated On:	2019-02-01
Status:	clientDeleteProhibited clientTransferProhibited clientUpdateProhibited
Name Servers:	ns-1.amazon.com ns-2.amazon.com ns-3.amazon.com

Registrant Contact

Name:	Hostmaster, Amazon Legal Dept.
Organization:	Amazon Technologies, Inc.
Street:	P.O. Box 8102
City:	Reno
State:	NV
Postal Code:	89507
Country:	US
Phone:	+1.2062664064
Fax:	+1.2062667010
Email:	hostmaster@amazon.com

↑ Screen capture of WHOIS (whois.com) information for the domain name relentless.com.

amazon.com

Updated 3 days ago

Domain Information

Domain:	amazon.com
Registrar:	MarkMonitor Inc.
Registered On:	1994-11-01
Expires On:	2024-10-30
Updated On:	2019-05-07
Status:	clientDeleteProhibited clientTransferProhibited clientUpdateProhibited serverDeleteProhibited serverTransferProhibited serverUpdateProhibited
Name Servers:	ns1.p31.dynect.net ns2.p31.dynect.net ns3.p31.dynect.net ns4.p31.dynect.net pdns1.ultradns.net pdns6.ultradns.co.uk

Registrant Contact

Name:	Hostmaster, Amazon Legal Dept.
Organization:	Amazon Technologies, Inc.
Street:	P.O. Box 8102
City:	Reno
State:	NV
Postal Code:	89507
Country:	US
Phone:	+1.2062664064
Fax:	+1.2062667010
Email:	hostmaster@amazon.com

Screen capture of WHOIS (whois.com) information for amazon.com. Note that the registration for relentless.com preceded the purchase of amazon.com by over a month.

Grifting the Amazon

Published by
Unknown Unknowns
New York, NY
September 2019

Design:
Anisa Suthayalai, Default

Special thanks:
Bill Jordan
Rachel Herschman
Amanda Noonan
Natalie Kerby

A version of this story
was shared in 2018
at Wordhack,
Babycastles Gallery
in New York
organized by
Todd Anderson.

ISBN-13: 978-0-9913923-4-6

unknownunknowns.org

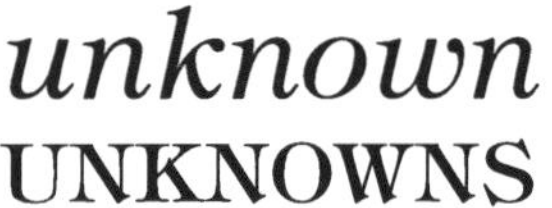